The Institute of Biology's
Studies in Biology no. 143

Animal Taxonomy

H. E. Goto

Senior Lecturer and Director of Undergraduate
Studies in the Division of Life Sciences,
Imperial College, University of London

Edward Arnold

First published 1982
by Edward Arnold (Publishers) Limited
41 Bedford Square, London WC1 3DQ

British Library Cataloguing in Publication Data

Goto, H. E.
    Animal taxonomy.—(The Institute of Biology's
    studies in biology; no. 143)
    1. Zoology—Classification
    I. Title   II. Series
    591'.012   QL351

    ISBN 0-7131-2847-X

Photoset and printed by Photobooks (Bristol) Limited

General Preface to the Series

Because it is no longer possible for one text book to cover the whole field of biology while remaining sufficiently up to date, the Institute of Biology proposed this series so that teachers and students can learn about significant developments. The enthusiastic acceptance of 'Studies in Biology' shows that the books are providing authoritative views of biological topics.

The features of the series include the attention given to methods, the selected list of books for further reading and, wherever possible, suggestions for practical work.

Readers' comments will be welcomed by the Education Officer of the Institute.

1982   Institute of Biology
                                       41 Queen's Gate
                                       London SW7 5HU

Preface

In recent years there has been a radical change in the study of taxonomy. It no longer deserves the reputation it still has in certain quarters, even among some professional biologists, of being a rather dry and static subject dealing with specimens preserved in museums. It is now an exciting field of research in which rapid advances are being made both in practice and theory. It employs techniques developed in the chemical, physical and mathematical sciences and takes regard of theoretical studies in evolution and classification. It is, in addition, of very great importance to biologists working in other fields for it is now recognized that precise identification may be vital in their work.

The subject is now so extensive that the coverage here has to be very selective. Some of the topics omitted are dealt with excellently in the works listed under Further Reading.

My very great thanks are due to Valerie Forge without whose assistance, typing and proof-reading this book would never have been completed. I should like to thank Professor R. G. Davies for his useful comments on Chapter 8; Dr C. Loumont (née Vigny) of Geneva for the loan of her original photographs of the sound spectrograms shown in Fig. 5–1, and to Drs C. A. Wright and D. R. Ragge of the British Museum (Natural History) for the special preparation of Figs. 3–5 and 5–2 respectively.

London, 1982   H.E.G.

Contents

1 Introduction

Modern taxonomy is a very broad subject encroaching on the theory and practice of many scientific disciplines. The rise of experimentation at the turn of the century gave rise to the temporary eclipse of studies in systematic biology and classification. However, during this period conflicting data arose from experiments on closely-related species in, for example, physiological, embryological and biochemical investigations. As a result the need for accurate identification became obvious. There was also a growing need to record, store, retrieve and pass on accurate information about the increasingly large number of species. This was undoubtedly one of the causes that led to a revival of interest in taxonomic studies.

The *New Systematics*, as it came to be known after the publication in 1940 of J. S. Huxley's book of that name, is quite different from the discipline as previously understood. It is now based to a large extent on the experimental sciences that once caused its eclipse. The new philosophy derived from the *Synthetic Theory* of evolution and the modern biological concept of the species also had a very significant effect on the subject. In the last 10–15 years enormous developments, both procedural and conceptual, have been made, especially in the areas of chemo-, immuno-, and numerical taxonomy.

Some indication of the size of the problem confronting a taxonomist may be seen from the following data. More than one million species of animals have been named; over 13 000 new species and nearly 1000 new subspecies are described in most years. In the same period about 2000 new genera and subgenera are erected, and some 24 000 nomenclatorial changes made at levels above the species. Of the new species over half are usually insects.

Ernst Mayr has estimated that about 99% of all birds are now known and that about 90% of mammals and reptiles have already been discovered. In many groups, however, the number is very much smaller: less than 10% of marine invertebrates, for example, have probably been described. Most of the 13 000 new species that are described each year belong to relatively unknown categories like marine invertebrates, but occasionally even new mammals are found. For instance, in 1973 a new marsupial mouse belonging to the family Dasyuridae was discovered in South Australia. Perhaps some 20 to 30 years may elapse before another new mammal will be found.

In general the number of different kinds of animals tends to increase, but in very well studied groups such as the birds there is actually a decrease. This is because named species when originally described may not represent true biological species. Instead they may simply be variants of one sort or another within a single species. It has also frequently occurred that different names have been given to the same species of animal. As these are discovered so the number of true species decreases. C. A. Wright (1971) has described how in 1758 one species of the African snail *Bulinus* was known, this was *B. senegalensis*. There remained only a single species in this genus until about 1830 when the number began to increase. By 1910 there were approximately 100 known species. Since then there has been an intensive study of the genus because some of its members were found to be important vectors of the parasitic worm *Schistosoma* that causes a great deal of human suffering in many tropical countries. As a result we now know that there are approximately only 30 true species.

Correct identification is vital in many biological fields, especially in ecology and various areas of applied zoology such as parasitology, medical and agricultural entomology, and in conservation. Two examples will suffice to demonstrate the importance of correct identification.

There is a complex of species of moths in the genus *Scirpophaga* whose larvae cause severe economic loss in some tropical and subtropical crops, particularly rice and sugar-cane, by boring into their stems. There are very few external characters that can be used to separate the species of *Scirpophaga* reliably. These moths are mostly white with few or no markings on the body or wings. When markings do occur they are very often unreliable for identification purposes. This situation has led to frequent misidentification and a great deal of confusion. Inappropriate control measures have resulted from incorrect identification. Studies by A. Lewvanich (1972) have revealed new characters particularly in the genitalia. These now enable identifications to be made with a very high degree of certainty. For example, it is now possible to separate *S. nivella*, a very important pest of rice from *S. excerptalis* that damages sugar-cane. The identification and biology, and therefore control measures, of these two species had hitherto been confused.

The second example is now a classic one. It was realized early this century that the distribution of malaria did not always coincide with that of the mosquito *Anopheles maculipennis*, its supposed vector in Europe. Even when the parasite (*Plasmodium*) was introduced into non-malarial areas where *A. maculipennis* was present, the disease was not transmitted. Such a situation occurred after the First World War when no cure was known for the disease and when troops returned home carrying the malaria parasite. This remained a mystery for some years until it was noticed in parts of Holland where malaria was endemic that the breeding sites of *A. maculipennis* differed in the malarial and non-malarial areas.

Table 1 Some differences between the species and subspecies of the *Anopheles maculipennis* complex.

Species/Subspecies	Oviposition Site	Hibernation	Carrier
A. maculipennis s. str.	cool running freshwater	yes	no
A. labranchiae	warm brackish water	no	serious
A. l. atroparvus	cool brackish water	no	slight
A. messeae	cool standing freshwater	yes	rarely
A. melanoon (= *subalpinus*)	standing freshwater (especially rice fields)	no	no
A. sacharovi (= *elutus*)	shallow standing freshwater or brackish water	no	serious

Where malaria occurred, the mosquito was found to breed in warm brackish water; where it was absent, *A. maculipennis* bred in cool, fresh water. The vector was believed to be a particular variety of the species and was named *labranchiae*. The non-carrier was called *messeae*.

An intensive study of the ecology of *A. maculipennis* (in its original broad sense) led to the discovery of five sibling (very closely-related and similar) species, and of one subspecies. The true *A. maculipennis* was found in fact to be a non-carrier. Each of these species, and the subspecies, were found to differ in their ecology, biology and in the degree to which they acted as a carrier of the malaria parasite (Table 1).

A. labranchiae is more southerly in its distribution than its subspecies, *atroparvus*, but where they overlap in central Italy they hybridize. In the F_1 generation the females are fertile but the males are usually sterile. It is for this reason that *atroparvus* is given subspecific rather than specific status.

In addition to the differences indicated in Table 1, we now know that each of these species and the subspecies differ in geographical distribution, behaviour, reproductive compatibility and fertility of hybrids. Marked differences are also found in the colour patterns on the egg and the fine structure of the egg floats. The patterns have been shown by scanning electron microscopy to result from different degrees of development of surface microtuberculation obscuring more or less of the darkly coloured inner layer of the egg shell.

Chromosome studies have revealed interspecific differences, mainly translocations and paracentric inversions (p. 48). These are especially noticeable in the banding patterns of the giant (polytene) chromosomes.

Similar species complexes are now well known in other groups of mosquitoes as well as in some quite unrelated organisms.

1.1   Definition of terms

It is necessary at the outset to define the basic terms used in taxonomic studies, i.e. *systematics, taxonomy, classification* and *nomenclature*. In the past, and still to a considerable extent today, the first two of these terms have been used rather loosely and synonymously to imply, in a broad sense, the science of classification. There is now a strong tendency to restrict the scope of these terms and to define them more precisely.

In 1952, R. F. Blackwelder and A. A. Boyden contributed a paper entitled 'The Nature of Systematics' to the first number of the journal *Systematic Zoology* in which they defined systematics as 'the entire field dealing with the kinds of animals, their distinction, classification and evolution'. SIMPSON, in his *Principles of Animal Taxonomy* (1961) preferred to extend the scope of the definition to include not only 'distinction and evolution' but 'any and all relationships' among living organisms. This broad and all-embracing attitude is the hallmark of modern systematics. He defined taxonomy as 'the theoretical study of classification, including its bases, principles, procedures and rules'. Classification he defined as 'the ordering of animals into groups or sets on the basis of their relationships, that is of association by contiguity, similarity or both'. Two types of classification are implied here: contiguity implies genetic or phyletic (phylogenetic) relationship. Similarity, on the other hand, implies only what is called a phenetic relationship.

Construction of classifications based on contiguity is attempted by those who try to attain a 'natural' or 'biological' system reflecting genetic affinity and, as far as possible, phylogeny. A truly phylogenetic classification is never really possible without extensive fossil evidence.

Classifications based on similarity alone probably approach very near to a phylogenetic one if a sufficiently large number of characters is taken into consideration. This is the method employed in numerical taxonomy.

Classification results in the construction of a *hierarchy* such as that shown in Table 2. The ranks or categories shown on the left indicate levels in the hierarchy. Organisms or groups of organisms at a particular level (i.e. in a particular category) are known as *taxa*. Some intermediate categories, such as superfamily and subfamily, and their corresponding taxa exist but these are not all shown in the figure.

Table 2   Classification of the St. Kilda Wren, *Troglodytes troglodytes hirtensis*, as an example of a taxonomic hierarchy.

Category	Taxon	Vernacular name
Kingdom	Animalia	animals
Subkingdom	Metazoa	multicellular animals
Phylum	Chordata	chordates
Subphylum	Vertebrata	backboned animals
Class	Aves	birds
Subclass	Neornithes	—
Superorder	Neognathae	—
Order	Passeriformes	perching birds
Family	Troglodytidae	wrens
Genus	*Troglodytes*	wrens
Species	*troglodytes*	common wren
Subspecies	*hirtensis*	St. Kilda wren

The terms category and taxon are frequently confused. Mayr defines a taxon as 'a taxonomic group of any rank sufficiently distinct to be worthy of being assigned to a definite category', i.e. the 'groups' or 'sets' of Simpson mentioned above. The limits defining taxa are entirely subjective in nature, except perhaps in the case of those in the categories of biological species and subspecies. The concept of a biological species is one of populations that are interbreeding or potentially interbreeding in the wild, and normally reproductively isolated from other similar groups by one or more of a number of types of barrier, see MAYR (1970).

The term taxon is employed in a somewhat different way by numerical taxonomists who use it to refer to a group defined by similarity only, i.e. the species is a *morphospecies* rather than a biological species.

1.2 Taxonomic procedure

The tasks of a practising taxonomist are usually divided into three phases:

1. *The analytical or alpha phase*   This includes the basic procedures of the systematic zoologist. These are: (*a*) the arrangement of animals into easily recognizable groups that are phenotypically more-or-less homogeneous. Such groups are known as *phena*. When first recognized at the specific level they may not even be morphospecies. They may conceal a number of species within a complex. Frequently different forms of a polymorphic species, races (or subspecies) of a polytypic species, males and females in a sexually dimorphic species, and even sometimes juveniles and adults of the same species fall into different phena. The taxonomist must unite such phena correctly into a single taxon. The term phenon is used differently by numerical taxonomists (see p. 55). (*b*) The description of new taxa. At this stage these are usually limited to a few criteria just sufficient to characterize the taxon and to distinguish it from other similar taxa. These primary descriptions must be adequate to enable subsequent identification. (*c*) The provision of suitable names in accordance with the rules of the *International Code of Zoological Nomenclature*. At this stage species are known as nominal species. The provision of universally accepted names is essential for accurate future reference.

2. *The synthetic or beta phase*   This stage, also referred to as macrotaxonomy, comprises mainly the ordering of species into categories in the hierarchy, i.e. the provision of a classification. At first this is constructed from the maximum number of available characters, i.e. it is a purely phenetic classification.

3. *The biological or gamma phase*   This phase is concerned mainly with the analysis of infraspecific variation leading to the study of the origin, evolution and determination of the biological species and its subspecies.

In the majority of groups of animals taxonomic studies have not yet progressed beyond the alpha and beta stages. It is only in some vertebrates, especially the birds, and a few insect orders such as Lepidoptera (butterflies and moths) that any real progress has been made in gamma taxonomy.

No consideration is given in this book to the basic procedures of taxonomic discrimination, description, key construction and nomenclature. These are excellently covered in MAYR (1969), PANKHURST (1978), JEFFREY (1973) and the *International Code of Zoological Nomenclature*.

Similarly, although the philosophy underlying the construction of classifications is currently a major one for discussion among taxonomists, no detailed consideration will be given here. The main point of dispute in this controversy is whether 'recency of common descent' should be the only criterion in classification. Acceptance of this *cladistic* view, as pointed out by Mayr would relate the crocodiles more closely to the birds than to other reptiles even though they are closer to the latter 'as far as the total gene composition is concerned'. No account is taken of the degree of divergence and associated radical genetic change. The other main theory is that of *Evolutionary Classification* championed by Mayr. According to this view one should consider not only cladistic relationships, but also phenetic and chronistic ones. The last two relate to degree and rate of divergence. For fuller discussions, the reader is referred to HENNIG (1966), MAYR (1969) and SNEATH and SOKAL (1973).

2 Taxonomic Characters

The basic procedure in any taxonomic study from the analytical through the synthetic to the biological phase always involves the consideration of a number of taxonomic characters. MAYR (1969) defined a taxonomic character as 'any attribute of a member of a taxon by which it differs from a member of a different taxon'. Interpreted in this way characters may be of any morphological kind, external, internal, macroscopic, microscopic or ultrastructural; they may be chemical, physiological, behavioural, developmental, ecological, geographical or any other kind of attribute of the organisms.

2.1 Selection of characters

In selecting taxonomic characters, certain precautions must always be borne in mind. They must not be too variable within a particular taxon whether this be due to phenotypic or genotypic cause. Size, colour and patterning are examples of attributes often falling into this category. For instance variation in size may be related to nutrition; colour to food difference, insolation or humidity and patterning may vary with age. This is not to say that these characters are always unsatisfactory. In the primitive insect genus *Entomobrya* (order Collembola), for example, colour patterning, although often very variable between individuals and at different stages in the life history, is sometimes the only character by which some species of the genus can at present be separated. Such *unreliable characters* are usually only revealed after a considerable knowledge of a particular taxon has been amassed. Inevitably, therefore, in the alpha phase of taxonomy many characters of this nature are employed and have later to be abandoned. Frequently a character may be reliable or 'good' in one group, but unreliable in another to which it may even be closely related. In other genera related to *Entomobrya*, for example, colour and patterning are relatively constant within a species and therefore constitute valuable taxonomic characters.

The different conditions exhibited by a single character such as brown or grey pelage in a mammal are known as the *character states*. Characters with states showing reduction of one sort or another as a result of evolutionary change are also considered to be rather poor. Reduction or total loss of eyes or wings are examples of such unreliable *regressive characters*. Both of these examples are known to have occurred a number of times independently in many groups. To link taxa because they have, for example, a reduced number of ommatidia in a compound eye might well lead to the creation of a polyphyletic group. Differences in the relative value of characters from 'good' to 'unreliable', and in all their intermediates, have given rise to the concept of *character weighting*. A 'good' character, according to some taxonomists, is one to which extra weight should be given in considerations concerning the construction of classifications, especially if these have any pretensions of reflecting phylogenetic relationships. Characters to which they would assign high weight are those with so-called high *information content*. Characters whose states are restricted to particular taxa and are of little intra-taxon variability come into this category. Further comments on weighting will be made under the heading of numerical taxonomy (pp. 53, 57). In addition to unreliable characters, there are also some that are totally unacceptable or *inadmissible*. An example of this is the use of characters that are known to be completely correlated and therefore should be regarded as one. Examples are seen in an insect's ability to fly and the presence of fully functional wings; albinism and the lack of normally present pigmentation; and the red colour of blood in vertebrates and the presence of haemoglobin. During the synthetic phase of taxonomy when species are arranged in hierarchical categories to form a classification rarely, if ever, are single characters employed even though they may exist in a number of possible states. This is also true, but to a lesser extent, in the construction of keys for identification. Procedures employing single characters are said to be *monothetic* and those using a number of different characters at the same time are known as *polythetic*. It is the latter that is an essential aspect of numerical taxonomy but it is by no means confined to it.

2.2   Types of taxonomic character

Taxonomic characters are of two main types: qualitative and quantitative. The former are sometimes difficult to express precisely, especially when employed in numerical taxonomy. However, this is not always the case for such characters are often discontinuous as in the A, B, O blood groups. Discontinuous qualitative characters are generally the result of unifactorial inheritance, that is to say they are controlled by a single or a few genes, the environment having little or no effect on their expression. Quantitative characters, on the other hand, are usually more easily expressed numerically but may provide considerable difficulty when their occurrence is more or less continuous such as in the stature of man. Continuous variation of this sort is usually associated with determination by a number of genes (polygenic or multifactorial inheritance). These characters are also frequently subject to considerable phenotypic (i.e. environment related) variation. The use of continuously variable quantitative characters involves the sub-division of the continuum into definable sub-units. In the case of size, for example, one may have a character state from 15–25, from 25–35 and from 35–45 mm. Divisions of this sort are never entirely satisfactory. Quantitative characters of another type, known as *meristic*, are much more useful as they depend on countable features which may differ precisely from one taxon to another as, for example, the number of setae on a well-definable area or part of an arthropod's body, or the number of spots on the wing cases of a ladybird. Even meristic characters vary within a single taxon and must be used with caution – the number of vertebrae and the number of fin-rays in some fish vary with the temperature of the water so that specimens taken along a north-south axis of distribution may provide a more or less continuous variation or *cline* in the numbers of these structures. Specimens of the Pacific herring (*Clupea pallasi*) possess high vertebral and fin-ray numbers off the coast of Alaska, both gradually decline as their distribution approaches southern Californian waters.

The sources of taxonomic characters used in modern studies are too numerous to be covered satisfactorily in this book and a selection only of some of the more important ones is provided in the following chapters. Before proceeding with them, however, brief mention will be made of two areas of investigation that are still in a comparatively early stage of development.

2.3 Electron microscopy in taxonomy

In recent years, since the innovation of the electron microscope, ultrastructure and fine structure have provided useful data. Scanning electron microscopy (SEM) is used by taxonomists to a much greater extent than transmission electron microscopy (TEM). The great advantage of SEM lies not only in the greater magnification and resolution that it provides, but also because it gives a depth of focus about three hundred times as great as that of light microscopy. Useful magnifications of up to about × 80 000 are possible and with biological

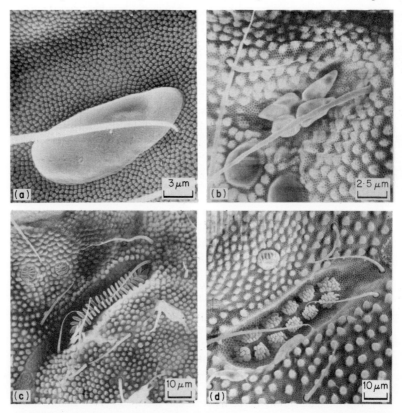

Fig. 2–1   Scanning electron micrographs postantennal organs of some Collembola: (a) *Folsomia candida*; (b) *Hypogastrura denticulata* (c) *Onychiurus flavidulus* and (d) *Onychiurus sinensis*.

material a resolution of about 12 nm can be obtained. Unlike TEM, scanning electron microscopy provides an almost three dimensional picture of the surface of a specimen. This is achieved by appropriate treatment of secondary electrons emitted from the material (usually gold) coating the specimen when bombarded by the beam of electrons scanning the surface of the structures under examination. The specimen can be moved and, to some extent, manipulated, whilst under observation on a television screen. Electron micrographs, such as those shown in Fig. 2–1 can be produced by photographing the image on the screen. Specific differences in fine structure can often be found in this way.

TEM studies are based on very thin sections cut with an ultramicrotome, 'stained' with electron opaque materials and viewed with a transmitted beam of electrons. Much greater magnifications can be achieved in this way than with either light microscopy or SEM.

TEM, although used to a lesser extent, also has its place in modern taxonomic studies. There is a genus of Protozoa, related to *Amoeba*, known as *Thecamoeba* whose species are not uncommon in damp leaf litter. They differ from *Amoeba* in having a shape more or less fixed by a pellicle and being without pseudopodia. They have been divided on the basis of their external appearance into two groups, those with smooth and those with rugose surfaces. The two groups have, by some authors, been placed not only in different genera but even in distinct suborders. Other studies based on features of the nuclear division have suggested another classification that does not coincide with that established on the smoothness or rugosity of the surface. F. C. Page and S. M. Blakey (1979) have examined sections of the surface with TEM. Their findings do not support the division of *Thecamoeba* according to either of these systems. They have found, however, that one of the species they studied, *T. granifera*, differed so markedly in surface structure and chemistry that they proposed it should be placed in a new genus *Dermamoeba*.

It is in groups such as the Protozoa where surface features appear to be few that TEM studies may be used to great value. Nevertheless they can play a very useful role in some other groups. Gremigni (1979) has studied the fine structure of cytoplasmic inclusions of the ovarian oocytes of planarian Turbellaria (free-living flatworms). His results support the classification and phylogenetic views put forward, on the basis of a number of criteria, by Ball (1974; 1977) and not those of other workers.

2.4 Recognition characters

The best types of specific characters are those used by the animals themselves to distinguish one another, especially in mate selection – the biological species concept being one of a naturally interbreeding population. The assumption is nearly always made that animals perceive each other with the same sensory limitations that we ourselves possess. It has been known for many years that not all animals are sensitive to the same wavelengths of light or hear the same frequencies of sound. It is only comparatively recently, however, that this knowledge has been put to use in taxonomy. Most vertebrates, especially those which are diurnal, are blind to ultraviolet light because it is absorbed in the lens of the eye. This is not true however of all animals. Among the arthropods both the simple ocelli and compound eyes are, at least in some cases, sensitive to UV radiation. The wing patterns of butterflies so familiar to us, may be seen quite differently by the insects themselves. It is now known that UV reflection patterns are present on the wings and that these sometimes bear little or no relation to those seen by humans. Such patterns have been used as taxonomic characters but at present only in a very limited number of taxa. Y. P. Nekrutenko (1964), for example, has employed UV reflection patterns in studies on the taxonomy of palaearctic species of the pierid butterfly genus *Gonepteryx* (brimstone butter-flies and their relatives) and more recently R. E. Silberglied (1972) has carried out similar studies on *Colias* (clouded yellow butterflies) in the same family. Polymorphism in these patterns has been observed, so has geographical variation and sexual dimorphism. We have a good example of the need for making taxonomic observations on as many individuals as possible within a given taxon. There is no doubt that, at least in some cases, the UV reflection patterns are used by the butterflies themselves in intraspecific recognition and can thus act as a prezygotic barrier to interspecific mating (see MAYR, 1970). This character therefore fulfils the requirement of the best type of specific character mentioned above.

In a similar way, ultrasonic sounds emitted in one way or another by a number of animals are now known to be used in both intra- and interspecific signalling. Taxonomic use of these sounds has been made to a very limited extent but may in the future give us characters of value equal to those provided by the 'hidden' UV reflecting patterns. In future studies UV patterns and ultrasonic sounds may be found to be of importance in the discovery of sibling species as well as in the separation of 'difficult' ones and also in the delimitation of higher taxa and their classification.

3 Chemotaxonomy

For many years it has been known that there are demonstrable differences in the biochemical composition of different organisms. The earliest to be noted were naturally those between higher taxa and not between species, genera or even families.

In 1871, E. R. Lankester wrote '*The chemical differences of different species and genera of animals and plants are certainly as significant for the history of their origin as the differences of form. If we could clearly grasp the difference of the molecular constitution and activities of different kinds of organisms, we should be able to form a clearer and better grounded judgement on the question how they have been developed, one from the other, than we now can from morphological considerations.*'

Since Lankester's time comparative biochemistry has progressed a long way and his prediction has proved to be well founded.

In 1958, F. H. C. Crick wrote that it could be argued that the amino acid sequences in the protein molecule '*are the most delicate expression possible of the phenotype of the organisms and that vast amounts of evolutionary information may be hidden away within them*'.

It is not surprising therefore that little chemotaxonomic work with animals has been carried out with substances other than protein, peptides and amino acids. There are, however, some exceptions. For example, there have been the studies on the distribution of different phosphagens, the indoles and imidazoles in primate blood and urine and bile salts.

Proteins are all very similar in chemical properties and it is difficult to isolate them by classical methods without denaturing them. For this reason the techniques most commonly used, in what is sometimes termed *protein taxonomy*, are those of chromatography, electrophoresis, immunology and various combinations of these. Included in such studies are conjugated proteins and enzymes. It is however the protein fraction of a conjugated molecule that nearly always provides the significant data (e.g. in haemoglobin studies).

Differences of two main kinds can be found in studies of protein taxonomy.

1. *Relative quantitative differences* between amounts of different proteins and, more important, of different constituent amino acids of the macromolecule. Such data are liable to vary according to the physiological state of the individual especially in relation to free amino acids in the cytoplasm and body fluids. Consequently the material must always be standardized as far as possible.

2. *Qualitative differences of three types exist*

(*a*) Different constituent amino acids – this is not so useful because most amino acids occurring in animal proteins can be present even in distantly-related or unrelated organisms, and very closely-related organisms may differ qualitatively in one or more amino acids if one studies, as one often does, not a complete protein molecule but a particular peptide. For example glycine is present at three positions in fibrinopeptide B in ox blood but is absent from that of some other artiodactyls, e.g. the dromedary camel where it is replaced by glutamine, arginine or aspartic acid.

(*b*) Of much greater significance is the relative order in which these amino acids are arranged along the length of the macromolecule. Studies on the structure of fibrinogen from different species have provided valuable information on mammal classification.

B. Blombäck and M. Blombäck (1968) studied the amino acid sequences in the fibrinopeptides split off from the parent fibrinogen molecule by the action of thrombin in 34 different species, mainly mammalian.

The fibrinogen molecule consists of three polypeptide chains (*a*, *β* and *γ*) linked by disulphide bridges (Fig. 3–1). The proteolytic enzyme thrombin splits off two fibronopeptides, one from the *a* and one from the *β* chain. It is in these two, usually known as fibrinopeptide A and B, that the sequences have been worked out.

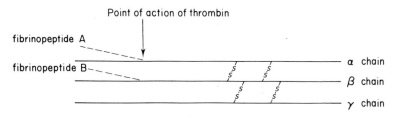

Fig. 3–1 Diagrammatic representation of a fibrinogen molecule showing formation of A and B fibrinopeptides.

Fibrinopeptides seem to be especially suitable for taxonomic studies as they constitute genetically equivalent parts of the fibrinogen from different species.

Many protein sequences have now been worked out (R. V. Eck and M. O. Dayhoff, 1966– ; L. R. Croft, 1973–).

The classification of species according to their amino acid sequences in the peptides agrees in general with accepted schemes based on morphology; and phylogenies, such as that illustrated in Fig. 3–2, can be constructed from the data so obtained.

Cytochrome C is present in cells of all living organisms. It contains some 110 amino acids. The sequence of these is now known for many organisms – our knowledge of these already fits in fairly well with current classifications.

(*c*) Variation in *shape* of the protein molecule. The chains of the macromolecule are held by weak electrical forces in certain definite shapes. These electrical forces are derived from hydrogen bonding. The shape of the macromolecule is important wherever it has to be matched or complemented to another molecule as in antibody formation and, for example, in the association between the prosthetic group (in this case haem) of the haemoglobin molecule with the protein fraction. Maintenance of the shape of the haemoglobin molecule is

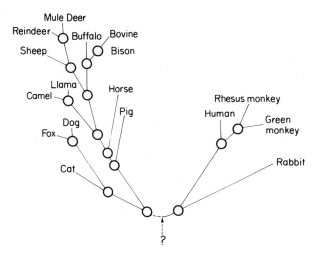

Fig. 3–2 Phylogenetic relationship of some mammals based on fibrinopeptide structure. (Modified from Eck, R. V. and Dayhoff, M. O. (1966). *Atlas of Protein Sequence and Structure*. National Biomedical Research Foundation, Maryland.

essential for proper functioning but provided that the correct shape is maintained many amino acid substitutions can be made – it has been estimated that of the approximately 150 amino acids making up the globin fraction of the molecule there were only nine sites where the same amino acid occurs in the haemoglobin of all the species examined. Molecular shape, therefore, although of some use may tend to be very stable in related taxa.

The techniques employed are numerous. Until recently some have been very time consuming and elaborate, such as those required for the determination of molecular shape and amino acid sequence. Since the introduction of automated techniques for amino acid detection, sequence studies have become much more rapid but they require very expensive and elaborate apparatus. There are, however, other analytical procedures that are simple and can be employed by a biologist relatively untrained in biochemical methods.

Two types of techniques are generally recognized. On the one hand there are the methods of comparative immunology that yield measures of overall similarity and, supposedly, of taxonomic affinity. Other techniques, such as those of chromatography and electrophoresis, can provide a number of characters in a single experimental procedure – these are called the *polyphenic methods*. They do not in themselves yield measures of affinity, but like morphological characters can of course, by appropriate mathematical procedures, indicate similarity and suggest affinity.

3.1 Chromatography

Chromatography includes a number of techniques for the separation of a mixture of solutes brought about by the differential movement of the individual solutes in a solvent moving through a porous medium. They provide a set of data (which in some cases can be quantified) on the presence or absence of individual chemical substances which, as mentioned above, do not give a similarity index in themselves but which can be made to do so.

Column chromatography is now little used in taxonomic studies. The most commonly employed form of the technique is partition chromatography either on paper or on a thin layer (TLC) of silica gel, alumina or other substance on a glass plate, the chromatoplate. In most instances, two-way chromatography is employed. This involves a second separation at right angles to the first run (Fig. 3–3 b and c).

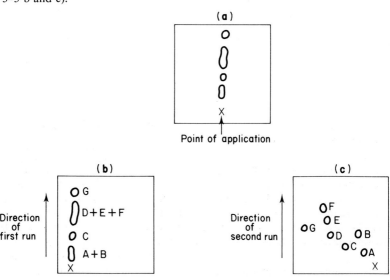

Fig. 3–3 Diagrammatic representation of single and two-way chromatography. **(a)** Single-way chromatogram, **(b)** two-way chromatogram after first run and **(c)** two-way chromatogram after second run.

With thin layer techniques the spots are very much more compact than is possible with paper chromatography and consequently smaller amounts of material can be separated. TLC also lends itself to preparative methods that permit confirmatory identification of the substances and their quantitative estimation.

The majority of chromatography in animal taxonomy has been carried out using the ninhydrin reaction to reveal the presence of amino acids by the blue-lilac colour that develops on heating. The spots can be identified by the comparative distances they have moved in both directions (Fig. 3–4).

At least 21 amino acids have been detected in homogenized adult mosquitoes. Most of them are present in all species but some interesting exceptions have been found. For example, aspartic acid seems to be absent in at least the nearctic species *Culex stigmatosoma*, and cysteic acid in *C. quinquefasciatus*.

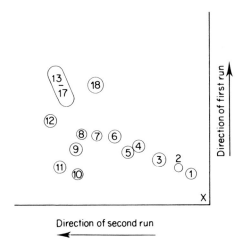

Fig. 3–4 Two-way chromatogram of a protein hydrolysate. Amino acids revealed by ninhydrin and identified by their positions resulting from distances of movement in both directions. x: point of application 1: cystine 2: aspartic acid 3: glutamic acid 4: serine 5: glycine 6: threonine 7: alanine 8: hydroxyproline 9: histidine 10: lysine 11: arginine 12: proline 13: methionine 14: valine 15: phenylalanine 16: isoleucine 17: leucine 18: tyrosine.

The use of total homogenates, often hydrolysed, have to be used with extreme caution as differences due to age, sex and physiological state are very liable to occur – they need very careful standardization and replication. The use of simpler or more purified substances, such as egg protein and mucus give better results.

Work on substances such as purines, pyrimidines and pteridines that fluoresce under ultraviolet light has also been carried out.

C. A. Wright (1959) has found that species-specific patterns in UV fluorescent chromatograms are obtainable with the mucus from British species of the gastropod *Lymnaea. L. stagnalis* and *L. truncatula* yielded virtually *no* fluorescent materials; *L. palustris* and *L. glabra* gave characteristic patterns unlike one another or any other species. *L. peregra* and *L. auricularia* showed a common basic pattern (incidentally shared with the African species *L. natalensis*). Within these common basic patterns there were distinctive species characters.

Subsequently (1964) C. A. Wright discovered the existence of a number of chromatographically distinct races of *L. peregra*. About 700 populations were sampled from all over the British Isles and five distinct basic pattern types have been identified each with characteristic subdivisions.

This is an important point to bear in mind when interpreting biochemical data because distinct subspecific and infra-subspecific differences are to be found in chemical as well as in morphological features. Herein lies one of the importances of replication.

Gas chromatography is a technique, based on column chromatography, in which the liquid solvent is replaced by a gas. This method is particularly useful for separating volatile materials but so far it has been little used in taxonomy. As indicated earlier, the best species specific character is the one that is employed by the animal itself for species recognition. In mammals and some other animals volatile scents are used for this purpose. Subjection of scent gland secretions to gas chromatography could possibly provide very useful taxonomic data.

3.2   Electrophoresis

Various methods of electrophoresis have, to a considerable extent, replaced those of chromatography. Electrophoresis is simply the movement of ions in solution under the influence of an electric field. Electrophoresis in simple solution (boundary or moving boundary electrophoresis) is now rarely used, having been replaced by *zone electrophoresis* – that is the method in which the solution containing the ions is supported in a more or less inert material such as starch, agar, cellulose acetate, polyacrylamide gel or paper. One of the most commonly used forms of zone electrophoresis is disc electrophoresis – so called because it uses a discontinuous buffer system. Proteins of similar mobility cannot normally be separated by either of these procedures. However, one of the latest techniques, that of polyacrylamide *gradient gel electrophoresis*, can separate substances of similar mobility but of dissimilar molecular size. In this method the proteins migrate through progressively smaller pores, the sizes of which can be regulated by adjusting the gel concentration. They are retarded and finally stopped where the pore size becomes too small to permit any further migration. The progressive reduction in pore size hinders the movement of the leading edge of migrating zones thus giving rise to compact concentrated zones. Substances of very similar size cannot be separated by this means. However, two-directional electrophoretic techniques may be used for this purpose. Substances of similar molecular size are separated along one axis. These are then further separated according to mobility by running the electrophoresis (zone) at right angles to the first.

Electrophoresis has been used at one time or another to separate practically all types of charged compounds but one of the most widespread uses is, as one might expect, for the separation of proteins, peptides and amino acids which contain both acidic and basic groups and have some of these groups in a charged or ionic form under normal circumstances.

The methods of electrophoresis have been used with considerable success on a number of different groups of animals.

C. A. Wright and G. C. Ross (1963), during their studies on the vectors of schistosomes, examined the electrophoretic patterns of blood proteins of planorbid snails. Like some other similar studies, these failed to give significant results because the protein composition of the blood changed both qualitatively and quantitatively during the life of the snail. Successful results were however obtained by using egg proteins which are unaffected by diet and which show some quantitative but not qualitative differences as the egg develops. Figure 3–5 shows egg proteins of *Bulinus* spp. separated on cellulose acetate strips.

They found that the basic patterns so obtained were characteristic of species groups and that there were minor differences within the patterns between species

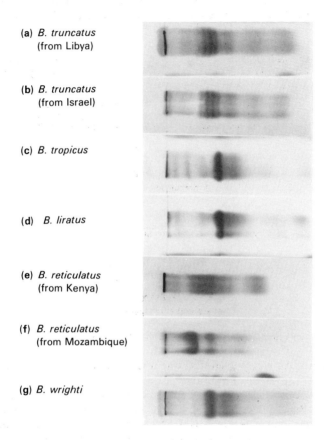

(a) *B. truncatus*
(from Libya)

(b) *B. truncatus*
(from Israel)

(c) *B. tropicus*

(d) *B. liratus*

(e) *B. reticulatus*
(from Kenya)

(f) *B. reticulatus*
(from Mozambique)

(g) *B. wrighti*

Fig. 3–5 Cellulose acetate electrophoresis of egg proteins of *Bulinus* (from WRIGHT, 1974).

and even between populations. The population specific patterns are genetically determined and breed true in the laboratory – this is really the ultimate test of the value of such data.

Another example of work on eggs has been carried out during the last 20 years by C. G. Sibley and his collaborators on egg-white proteins of well over 100 species of birds.

Their work is particularly interesting because birds are structurally so remarkably similar that there is often considerable difficulty in classifying them. One or two cases of this have been indicated in dealing with behavioural data (pp. 38, 39).

The Phoenicopteridae or flamingos, which have presented a considerable problem in recent classifications of birds, provides a good example. The flamingos resemble the Anseriformes (ducks and geese) in their bill structure, webbed feet, swimming behaviour and in their mallophagan parasites. On the other hand they share most of their anatomical features with the storks and herons (Ciconiiformes) but their cestode parasites suggest a link with the plovers and sandpipers (Charadriiformes) (Fig. 3–6).

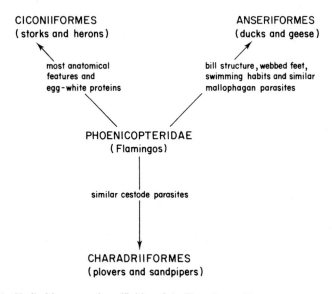

Fig. 3–6   Similarities suggesting affinities of the Phoenicopteridae.

This provides a very interesting case of conflict of evidence from different sources.

In a paper on the unsolved problems of avian systematics E. Stresemann (1959) placed the flamingos in a separate order but stated his belief in their ciconiiform affinities. E. Mayr and D. Amadon (1951), in their classification of recent birds, also assigned them to a separate order somewhere between the Ciconiiformes and the Anseriformes. In the revised classification of the birds of the World by A. Wetmore (1960) they are given subordinal rank within the Ciconiiformes.

This was the situation when Sibley published his early work on the egg-white proteins in 1960.

Sibley used paper-strip electrophoresis and a photo-electric density-measuring scanner to provide a pen-drawn curve or profile of the optical densities of the protein dyed with bromphenol blue after electrophoresis.

Comparison of the profiles for the lesser flamingo (*Phoeniconaias minor*) with those for a typical duck (*Anas georgica*) and a typical heron (*Ardea herodias*) are shown in Fig. 3–7. Note that as electrophoresis proceeds, the similarity between the profiles for the flamingo and the heron remains, but gradually diverge from that for the duck.

From such data Sibley was able to construct a hypothetical dendrogram of relationships. Note the positions of the ducks (Anatidae), flamingos (Phoenicop-

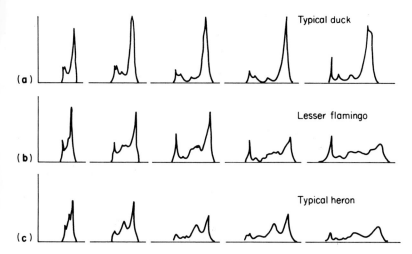

Fig. 3–7 Progressive density profiles for comparison of the lesser flamingo (**b**) with a typical duck (**a**) and a typical heron (**c**) (Redrawn after Sibley, G. C., 1960, *Ibis*, **102**, 215–84.)

teri), herons (Ardeidae) and the plovers and sandpipers (Charadriidae) (Fig. 3–8).

Subsequently Sibley and his co-workers have used other electrophoretic techniques. For example, thin layer electrophoresis of tryptic peptides derived by hydrolysis of the ovalbumins by the enzyme trypsin. Figure 3–9 shows the thin layer electrophoretograms for ciconiiforms, phoenicopterids and anseriforms.

Electrophoretic studies have also been carried out with enzymes, often with starch gel as a base.

The enzymes are extracted, usually by preparing a homogenate of the animal or organ to extract the enzymes from the tissues. The homogenate is then placed at one end of the gel and a low current passed through the gel for several hours. The gel is then incubated with a substrate appropriate to the enzyme to be studied together with a marker dye.

Experiments of this kind usually reveal that there is a small number of different forms of the enzyme that are distinct on the plate because of different speeds of movement. These all have similar enzyme activity but probably vary in the non-active part of the enzyme. They may differ in some activities, such as the rate of their catalytic reactions. Such forms are known as iso-enzymes or isozymes. Electrophoretic patterns of isoenzymes have proved to be very useful in taxonomic studies. C. A. Wright and his co-workers have investigated the digestive gland esterases of planorbid snails.

Esterases from the liver of various species of ducks have been studied and from the thoracic muscle of species of *Bombus* (bumble bees). Studies have also been made with enzymes from the flight muscles of bumble and honey-bees – in these cases the enzyme was glycerophosphate dehydrogenase. Many isoenzyme

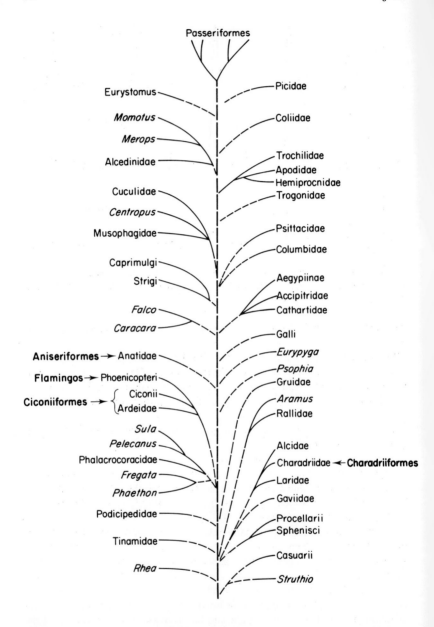

Fig. 3–8 Some relationships suggested by egg-white protein electrophoresis. Solid lines indicate reasonably certain relationships. Dashed lines indicate possible relationships. Gaps indicate unknown relationships (after Sibley, 1960).

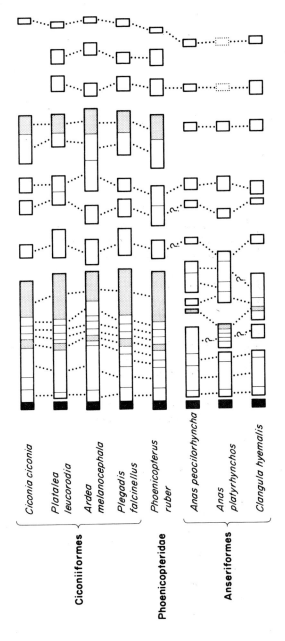

Fig. 3-9 Thin-layer electrophoretic patterns of the tryptic peptides of ovalbumins. Application point on left and cathode to the right. Dotted lines connect presumably homologous peptides. Different densities approximate to differences in the intensity of the ninhydrin staining reaction. (Modified after Sibley, *et al.,* 1969, *Condor,* **71**, 155–79.)

studies have been made with lactate dehydrogenase, for example, from the serum of various mammals, and from muscle and eye extracts of various fish (Fig. 3–10).

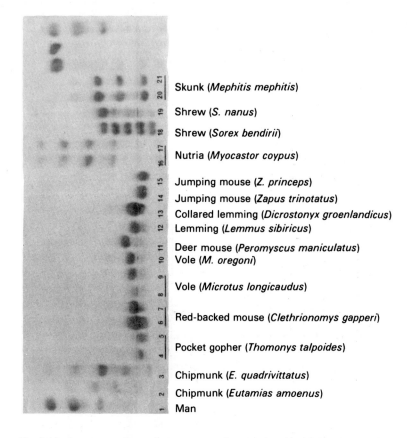

Skunk (*Mephitis mephitis*)

Shrew (*S. nanus*)

Shrew (*Sorex bendirii*)

Nutria (*Myocastor coypus*)

Jumping mouse (*Z. princeps*)
Jumping mouse (*Zapus trinotatus*)
Collared lemming (*Dicrostonyx groenlandicus*)
Lemming (*Lemmus sibiricus*)

Deer mouse (*Peromyscus maniculatus*)
Vole (*M. oregoni*)

Vole (*Microtus longicaudus*)

Red-backed mouse (*Clethrionomys gapperi*)

Pocket gopher (*Thomonys talpoides*)

Chipmunk (*E. quadrivittatus*)
Chipmunk (*Eutamias amoenus*)
Man

Fig. 3–10 Isoenzyme electrophoretograms of serum lactate dehydrogenase. (From WRIGHT, 1974.)

As in so many taxonomic studies, care must be taken in work with isoenzymes to avoid the assumption of polymorphism too readily because, for example, changes in isoenzyme patterns are known to occur during ontogeny of both birds and mammals (P. F. Fottrell, 1967). There is also strong evidence that certain environmental factors may affect lactate dehydrogenase patterns.

Two-way separations can be carried out with horizontal plates or on paper. Gradient gel electrophoresis is also employed for the separation of isoenzymes.

Electrophoretic studies of taxonomic interest have now been carried out on a number of substances in addition to those mentioned above, e.g. milk, haemolymph, tears, snake venom, muscle and liver. For some account of this see: LEONE (1964) and WRIGHT (1974).

The use of electrophoresis in combination with serological techniques will be discussed later.

An electrophoretic method that has recently been introduced into taxonomic studies is that of *isoelectric focusing*. In this technique the mixture of proteins moves electrophoretically through a medium, usually a polyacrylamide gel, in which a pH gradient has been established. The proteins come to a halt at the isoelectric point (pI) when there is a zero net charge on the molecule.

The gradient is established by subjecting a mixture of ampholites (aliphatic polyamino-polycarboxylic acids) to an electric field in which they move according to their net charge; those with a low isoelectric point (i.e. acidic ones) will move towards the anode and those with high pI towards the cathode. In this way a series of ampholites will arrange themselves between the negative and positive ends of the gel. They will then be stationary and, because of their high buffering ability, will impart a pH to the area of gel surrounding them related to isoelectric points of the particular ampholite. Different ranges of pH can be set up by the use of various ampholites. This technique gives very sharp bands of protein separation.

Electrophoretic mobility, in terms of migration speed, and heat stability in, for example, lactate dehydrogenase, have also provided useful confirmatory information in vertebrate phylogeny. The primitive conditions in fish are those of thermal instability and high mobility. This is maintained in the amphibians and lower reptiles. In mammals thermal instability remains and the mobility is increased. In higher reptiles and palaeognathous birds (ostrich, emu and kiwi) mobility remains high but the enzyme has become heat stable. Thermal stability persists in the majority of birds (i.e. neognathous species) but mobility is reduced. Enzymes exhibiting thermal instability only do so at relatively high temperatures – above those that are normally encountered in the organisms' environment.

3.3   DNA hybridization

Another technique, of an entirely different nature, that will probably provide taxonomic information of a very fundamental kind is that of DNA hybridization.

Comparison of DNA from different species gives a good indication of genetic and perhaps also of phyletic relationship. Sequence analysis of DNA by chemical means is very time consuming whereas DNA hybridization, although not providing as much information, is comparatively simple and quick.

The technique consists essentially of separating the two strands of the double helix and allowing single-stranded DNA from another taxon to come in contact and unite with it. The extent to which this occurs (i.e. the degree of hybridization) is an indication of the amount of identity along the molecule.

The procedure is carried out as follows: the solution containing the double-stranded DNA molecule of one species (the homologous organism) is heated to about 100°C for a few minutes. This results in the separation into two single polynucleotide strands. In order to prevent reformation of the duplex molecule, the solution is added to hot agar which is rapidly cooled to immobilize the strands.

DNA from a second species (the heterologous one) is labelled with C^{14} or P^{32}. The radioactive molecule is then fragmented mechanically into relatively short polynucleotide segments about 1000 nucleotides in length. These are still double-stranded and are heated to separate the strands and rapidly cooled to prevent them joining again. The resultant short single strands are immediately added to agar in which the homologous DNA is embedded. On re-heating to 60°C, the fragments unite with complementary base sequences of the homologous strands. The remaining single-stranded fragments are washed out of the agar with salt solution. The degree of radioactivity remaining gives an indication of the amount of reformation of the double helix, i.e. of the genetic affinity between homologous and heterologous organisms.

4 Immunotaxonomy

Some substances of high molecular weight, such as proteins, nucleic acids, polysaccharides, lipids and various combinations of these, when injected into an animal, will stimulate the production by the lymphatic tissue of *antibodies*. The injected substances are known as *antigens*.

When a solution of antigen is mixed, in correct proportions, with serum containing antibodies (*immune serum* or *antiserum*) precipitation occurs. This is known as the *precipitin reaction*. Such reactions between antibodies and the *specific antigen*, that originally stimulated their formation, are known as *homologous* or *reference reactions*. Precipitation may also take place, but to a lesser extent, if the antigen is chemically different from, but related to the original one. Reactions of this type are said to be *heterologous* or *cross reactions*.

In general, the more closely-related animals are to one another, the more similar will be their chemical make-up and consequently also any antigens derived from them. In heterologous reactions, the closer (taxonomically) are the animals which produced the antigens, the greater will be the degree of precipitation. This is the *Law of General Proportionality* which is the basis of immunotaxonomy or serotaxonomy.

A single protein or other antigenic material is now known to have not one but many *antigenic sites* or *determinants* (due to different amino acid sequences) which can bind them to antibodies. This gives them a type of multivalency. The more closely-related are the antigens, the more determinants they have in common.

Antibodies are serum proteins of a type known as *immunoglobulins* or *γ-globulins*. Their molecules each consist of two heavy (long) and two light (short) polypeptide chains linked by disulphide bonds (Fig. 4–1a).

The points of reaction with the antigenic determinants are at the free NH_2-terminal ends of the light and heavy chains giving a bivalency to the

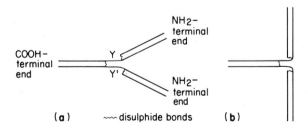

(a)   ⁓ disulphide bonds   (b)

Fig. 4–1 Diagram illustrating the basic structure of antibodies (**a**) and (**b**) show different positions of rotation of the arms of the polypeptide chains.

molecule. Bonding to the antigen is facilitated by possible rotation up to 90° at points Y and Y' (Fig. 4–1b). For further details and references see INCHLEY (1981).

Linking of bivalent antibodies to multivalent antigens gives rise to three-dimensional lattices which coalesce to form a precipitate. It is for this reason that a decrease in the number of suitable antigenic determinants (or lowering of valency) in heterologous reactions results in lowered precipitation according to the *Law of General Proportionality*.

For practical purposes, rabbits are commonly used as experimental animals for the production of antisera. Various sources of antigen have been used in immunotaxonomy, e.g. egg proteins, serum proteins, eye lens proteins, tissue homogenates and even total animal homogenates (in organisms of very small size). The use of homogenates and unpurified serum or other protein mixtures has certain disadvantages but may still provide useful immunotaxonomic information.

The most commonly used immunological procedures in taxonomic studies are:

i)   Serial dilution techniques.
ii) Ouchterlony and other gel diffusion methods.
iii) Immunoelectrophoresis.

4.1   Serial dilution techniques

These methods are based on the fact that there is an optimal combining ratio between antigen and antibody that results in maximum precipitation on either side of which precipitation falls off. This is because the formation of antigen/antibody lattices is reversible so the precipitate may redissolve if either excess antibody or antigen is present. In practice it is usual to use a constant concentration of antibody and to vary the amount of antigen (Fig. 4–2a).

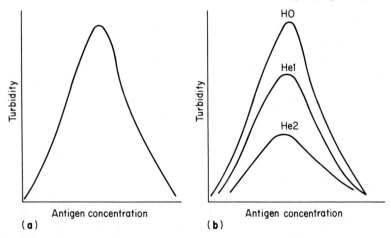

Fig. 4–2   (a) Relation between amount of precipitation and antigen concentration. (b) Comparison of precipitation curves for homologous (Ho) and heterologous (He1 and He2) reactions.

In comparison of curves for the homologous with various heterologous reactions it is found that the more distant the relationship the flatter the curves (Fig. 4–2b).

Various methods can be used for measuring the amount of precipitation in order to produce these *complete titration curves*. It may, for example, be centrifuged and weighed but a much simpler and faster method is, however, more commonly employed. This is by use of the Boyden technique which determines the degree of precipitation by a photoelectric device for estimating the turbidity of the solution.

A simple example of this technique is that provided by Boyden's comparison of the complete titration curves for precipitin reactions between horse serum (homologous antigen), the sera of the mule (he-ass × mare), hinny (she-ass × stallion) and ass (heterologous antigens), and anti-horse serum (antibody) (Fig. 4–3). The anti-horse serum is produced by injecting horse serum into rabbit. As one would expect, curves for the hybrids are intermediate between those of the horse and ass.

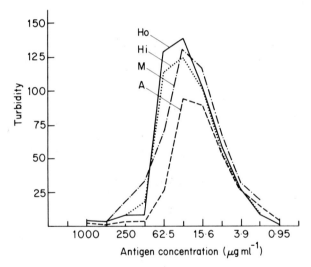

Fig. 4–3 Some equine precipitation curves (after Boyden). Ho, horse; Hi, hinny; M, mule; A, ass. Turbidity values are galvanometric readings.

Comparison of curves for different animals can be made quantitatively by calculating the areas beneath the heterologous reaction curves and expressing them as percentages of the area beneath the curve for the homologous reaction.

The data derived in this way will only indicate the distance between each individual animal used in the experiment and the single animal providing the antigen for the reference reaction. It is of greater value to be able to envisage the interrelationships as a three-dimensional model with each taxon placed at the correct distance from all others. This is made possible by employing each of the animals in turn as the reference or homologous taxon.

Studies of this sort have been carried out (A. A. Boyden, 1943) on four families of crabs: Cancridae, Goneplacidae, Portunidae and Xanthidae (Fig. 4–4)

The numerals represent the relative distances separating the families, each calculated from the average figure for all the species studied.

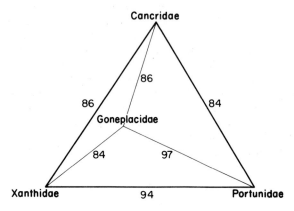

Fig. 4–4   Pyramidal representation of immunological distances between four families of crabs (after Boyden).

4.2   Ouchterlony and other gel diffusion methods

Complete titration methods require comparatively large amounts of antigen and antibody so the much more economical methods of gel diffusion are now commonly used. In these techniques the antigen and antibody are allowed to diffuse through the gel, usually agar. Where they meet, a line of precipitation occurs.

The most popular of these techniques is the *double diffusion method* of Ouchterlony. In this method the gel is placed in a Petri dish and circular wells cut in the agar, one in the centre to contain the antibody surrounded by about six others for the antigens from different animals (Fig. 4–5a).

The antibody diffuses outwards towards the wells containing the antigen, and when they meet the latter diffusing inwards, opaque lines or arcs of immunoprecipitation become visible (Fig. 4–5b). An example of the use of Ouchterlony gel diffusion may be seen in C. A. Wright's demonstration of the close relationship of the gastropod snail *Bulinus obtusispira* to the *africanus* group of species.

Precipitation arcs act as barriers to further diffusion of the reactants but they spread lengthwise until they meet the arcs from adjacent antigen-antibody reactions(Fig. 4–5c). Where adjacent wells contain identical antigens (i.e. from taxonomically identical animals) the arcs become continuous (*i*). Where there is partial identity a confluence of arcs may occur but with a peripherally projecting spur (*ii*). In the case of non-identity, the arcs do not join but cross one another (*iii*). The cause of the total confluence in (*i*) is the existence of the diffusion barrier set up by the precipitate and operating against the reactants giving rise to it. Different reactants, however, may pass through completely so that the arcs

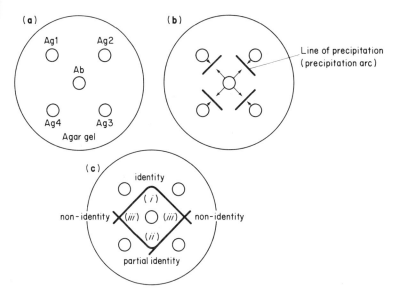

Fig. 4-5 Diagrammatic representations of the double diffusion technique (**a**) Ouchter-lony plate with antibody (Ab) and antigens (Ag1–4) in wells. (**b**) precipitation arcs forming where the diffusing antibody and antigens meet at optimal concentrations, (**c**) Ouchterlony spectrum showing the three main types of information derivable.

will continue to grow in length and will cross one another as in (*iii*). In the case of partial identity, i.e. where there are, for example, a common determinant and two unshared ones the common determinant will react with the corresponding antibody (Ab1) and set up a barrier but will let through the antibody (Ab2) to the unshared determinant. These will precipitate on the outer side of the barrier giving rise to the spur.

4.3   Immunoelectrophoresis

In such multiple systems where there are mixtures of antigens in a single well and corresponding mixtures of antibodies, it often becomes difficult to interpret the arcs correctly. It is here that immunoelectrophoresis is of value.

The antigens are first separated on the basis of their electrophoretic mobility and then are allowed to diffuse through the gel towards an advancing line of diffusing antibody. Where they meet, separated precipitation arcs occur.

Immunoelectrophoresis is now often carried out on a microscale, frequently in a gel coating one surface of a microscope slide (Fig. 4–6). The gel may be agar, starch, polyacrylamide or some other suitable material.

Some other immunological techniques, such as *complement fixation* and *microcomplement fixation*, have recently been introduced into taxonomy but their use is, as yet, not sufficiently widespread to describe here. An up-to-date review of techniques, application and results may be found in WRIGHT (1974).

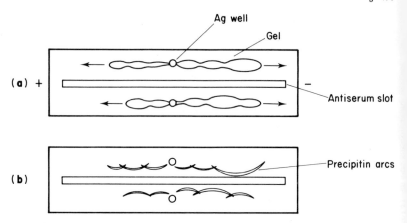

Fig. 4–6 Comparison of precipitin arcs from two different taxa by immunoelectrophoresis (**a**) electrophoretic separation of antigen components (**b**) precipitin arcs formed where diffusing antigens and antibody meet.

4.4 Findings from immunological techniques

The use of immunological techniques in animal taxonomy is now fairly widespread. They have provided information under the following headings:

4.4.1 *Determination of relationships of major groups*

A. A. Boyden and D. Gemeroy (1950) have shown that the Cetacea (whales, dolphins and porpoises) are more closely related to the Artiodactyla (even-toed ungulates) than to any other mammals; P. A. Moody and D. E. Doniger (1956) have demonstrated the lack of close relationship between the duplicidentates (rabbits and hares) and other rodents, thus giving support to the view that the former should be separated from the true rodents as the order Lagomorpha.

4.4.2 *Determination of the taxonomic position of individual genera or species*

C. A. Leone and A. L. Wiens (1956) have provided evidence that the giant panda (*Ailuropoda*) has closer affinity with the bears (Ursidae) than to the small red panda which is probably related to the raccoons and kinkajous (Procyonidae); the North American musk ox (*Ovibos*) is immunologically closer to the sheep and goats (Caprinae) than to the cattle (Bovinae) as demonstrated by P. A. Moody (1958). The mammalian order Edentata classically included the armadillo (*Dasypus*), the giant pangolin (*Manis*) and the aardvark (*Orycteropus*). On morphological and palaeontological grounds, G. G. Simpson (1945) suggested that they should more correctly be placed in separate orders: Edentata *sensu stricto* (*Dasypus*), Pholidota (*Manis*) and Tubulidentata (*Orycteropus*). Immunotaxonomic work by various authors has provided supporting evidence for this view and has also indicated that the Pholidota and Tubulidentata are more closely related to one another than either is to the Edentata *s. str.*

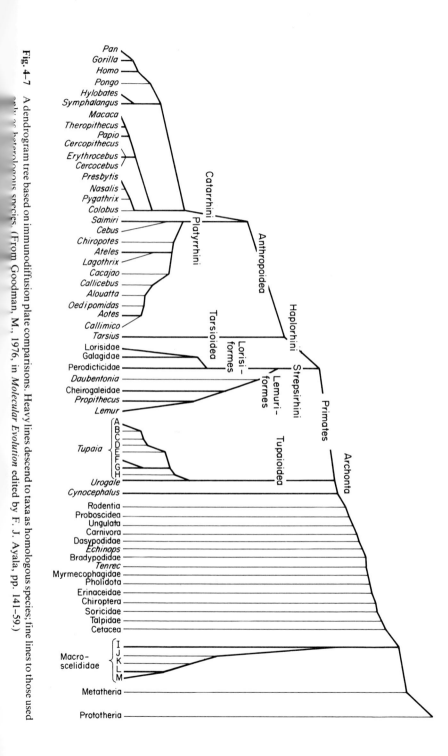

Fig. 4-7 A dendrogram tree based on immunodiffusion plate comparisons. Heavy lines descend to taxa as homologous species; fine lines to those used early as heterologous species. (From Goodman, M., 1976, in *Molecular Evolution* edited by F. J. Ayala, pp. 141–59.)

4.4.3 Determination of interrelationships of species within a group

Reference has already been made to the studies of Boyden on crabs (see also C. A. Leone, 1949). Similar studies have been made, for example, on turtles by W. Frair (1964), on cockroaches by W. P. Stephen and J. H. Cheledin (1970) and on species of the pigeon genus *Columba* by Irwin and others in a series of papers from 1932 onwards.

4.4.4 Confirmation of phylogenetic speculations

M. Goodman (1976) has illustrated this use of immunodiffusion with special reference to primates and their affinities (Fig. 4–7). W. Manski *et al.* (1967) constructed a similar phylogenetic tree based on immunoelectrophoretic studies of vertebrate eye-lens proteins.

In addition, the existence of geographical races and sibling species have been discovered by immunotaxonomic studies.

5 Behavioural Characters in Taxonomy

Behavioural characters have not as yet been used to any great extent in animal taxonomy, largely because very few comparative studies have been undertaken involving more than a few species within a single taxon. There is, however, great potential in their use. E. Mayr (1963) has pointed out that they are 'often clearly superior to morphological characters in the study of closely-related species, particularly sibling species'.

5.1 Sound

The use of such characters is by no means new. Aldrovandi, the 16th century naturalist of Bologna, based his classification of birds partly on behaviour, and Gilbert White, the 18th century parson and naturalist of Selborne, distinguished clearly for the first time the three British leaf warblers: the chiffchaff (*Phylloscopus collybita*), the willow warbler (*P. trochilus*) and the wood warbler (*P. sibilatrix*). This he did using behavioural criteria, especially their song patterns. These are sympatric birds of very similar appearance in which vocalization is used in species recognition and the establishment of territoriality. Characters, such as this, which provide species recognition by the animals themselves are of very high value in taxonomy.

Difference in song is not confined to sympatric organisms. For example, the very closely-related and similar allopatric South Island kiwi (*Apteryx australis australis*) and North Island kiwi (*A. australis mantelli*) of New Zealand have quite distinct song patterns.

These are, of course, completely isolated being apterous and allopatric and do not need to use vocalization as an isolating mechanism or for species recognition.

Songs which do provide an isolating mechanism in closely-related sympatric species have evolved as a result of strong selection pressure. The difference may be so great as to have masked any similarity due to the natural affinity of the birds. There would be no selective advantage in retaining similarity of song, for example, in whole genera, in fact the opposite may be true. For this reason vocalization in birds may be useful in distinguishing closely-related species but of little or no use in taxonomic studies of higher taxa at, for example, the family or ordinal level. At this level there may be gross differences due, in some cases, to morphological features such as the characteristic 'quack' of ducks.

There are, however, a few exceptions to this generalization. J. Delacour (1943) has described how the three main groups of estrildine birds of the family Phoceidae (the grass finches, the waxbills and the manakins) each have characteristic types of songs common to their group but quite distinct between groups (CATCHPOLE, 1979).

Sub-specific song variations, as in *Apteryx*, is fairly common so also is that between local populations. Individual variation is also well known in many birds, e.g. the blackbird (*Turdus merula*). Marked individual variation in vocalization is usually associated with the establishment of territory. Care must be taken in taxonomic studies to bear the possibility of local and individual variation in mind.

Comparison of bird song is usually made with the aid of sound spectrograms or Sonagrams. These provide a visual representation in the form of a permanent record of vocalization, recording differences in frequency, amplitude and duration of the various components of the song. The basic display is one of frequency/time rather than amplitude time but indications of relative amplitude (loudness) are shown by the level of intensity of darkness of the final graphic recording. The frequency, up to 16 kHz is indicated on the vertical axis and time on the horizontal.

The song is usually recorded, in the first instance, on tape and then transferred electrically to the sound spectrograph where it is re-recorded on to a magnetic loop. At subsequent revolutions of the latter the recording is scanned by a narrow band analysing filter usually of 45- or 300-cycle band width, recording

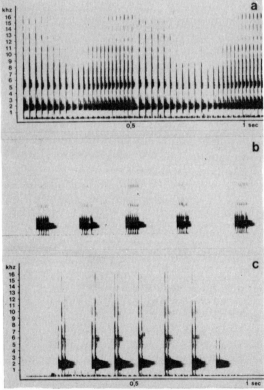

Fig. 5–1 Sound spectrograms (Sonagram) of the mating calls of three subspecies of the toad *Xenopus laevis* (**a**) *X.l. laevis*, (**b**) *X.l. petersi* and (**c**) *X.l. victorianus*.

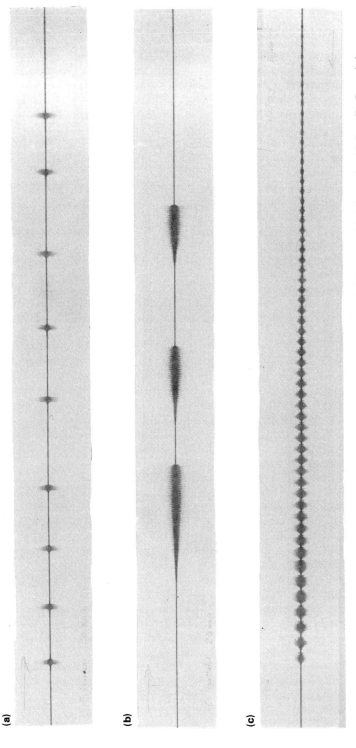

Fig. 5-2 Comparison of oscilloscope recordings of the songs of three species of *Chorthippus*: (a) *C. brunneus*, (b) *C. biguttulus*, (c) *C. mollis*. Recorded at a paper speed of 20 mm s^{-1}. (From recordings made by Dr D. R. Ragge.)

first the lower frequencies and progressively higher frequencies at each subsequent revolution. A paper-covered drum rotates synchronously with the magnetic loop and the filter output is recorded on it by a stylus that rises together with the scanner. There is a maximum frequency accuracy of 45 Hz and a time resolution of 1.5 msec. Modifications of the basic apparatus can give a more precise measure of amplitude.

Visual examination of the sound spectrograms provides a very simple and decisive method of comparing the sounds of amphibia (Fig. 5–1), birds and other animals.

The use of sound patterns in animal taxonomy is not confined to birds. There are many examples in other groups, e.g. Amphibia, Mammalia (including ultrasonic sounds in rodents) and especially in many groups of insects e.g. cicadas (and other Hemiptera), and Orthoptera. T. J. Walker (1964) has estimated that about 50% of North American Gryllidae (crickets) and Tettigoniidae (bush crickets) were discovered by the differences in the song patterns of sibling species. Among the Acrididae (short-horned grasshoppers) we find an excellent example of the taxonomic value of sound patterns: *Chorthippus brunneus, C. biguttulus* and *C. mollis* are very similar in colour and morphology, yet they have quite dissimilar sound patterns as shown in the oscilloscope recordings in Fig. 5–2.

Among the Diptera (two-winged flies) studies have been made of the sound produced by the wing-beat of male *Drosophila* during courtship display. The males extend one wing (sometimes both) and vibrate it whilst pursuing the female. If she is of the same species, is mature and unmated, copulation occurs. This seems at first sight to be a simple visual display which acts as a species recognition signal at the time of mating and also as an ethological prezygotic barrier to hybridization. However, H. T. Spieth (1952) has found, in a study of about 40 species that many have apparently identical displays. As these species successfully mated with individuals of the same species in complete darkness, the signal must be more than simply visual. It is now known that the sound resulting from the single wing beat differs in frequency, sound interval and duration of note. *Drosophila melanogaster* and its sibling *D. simulans*, for example both produce sounds of about 330 Hz at 25°C but the former at a rate of 29 per second and the latter at only 20 per second. The more distantly related *D. pseudoobscura* produces sounds of a higher frequency (about 500 Hz) but at a rate of only five per second.

It is more usual to employ a recording oscilloscope rather than a Sonogram in the study of insect sounds. The oscilloscope gives a precise amplitude/time display and both parameters may be adjusted over wide ranges. Values for frequency can also often be obtained by measuring the interval between cycles. If harmonics are present, masking may occur of the lowest, fundamental frequency, that gives the sound its pitch.

Sound spectrograms are more useful where there are sounds of different frequencies produced simultaneously in the song – these can be separately recorded as mentioned above. The oscilloscope, on the other hand, is employed where only a single frequency is involved.

5.2   Bioluminescence

Patterns produced by bioluminescence also have their place in taxonomy. H. S. Barber (1951) studied the flashing behaviour of species of the eastern North American fireflies of the genus *Photuris* (Coleoptera: Lampyridae). He discovered that they differ in frequency, intensity, colour and 'shape' of individual flashes or flash sequences. These may be of short or long duration, they may be of constant intensity or may either increase or decrease in intensity (Fig. 5–3).

At the commencement of these studies three species of *Photuris* were known; 18 sibling species were discovered as a result of them. Since then other differences have come to light, e.g. differences in preferred habitat, in breeding season and in body colour (slight only).

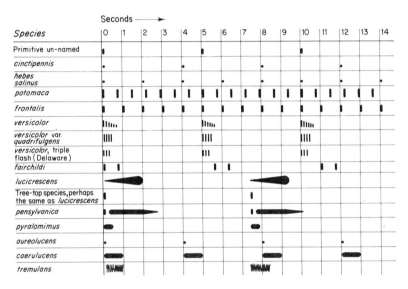

Fig. 5–3 Flash patterns of *Photuris* spp. Height indicates intensity of the flashes. (From MAYR, 1969, after Barber, H. S., 1951, *Smithsonian Miscellaneous Collection*; **117**, 1–58.)

5.3   Activity

Various types of activity pattern have also been employed by taxonomists. For example, J. Crane (1941) undertook a comparative study of the behaviour of the fiddler crab (*Uca*) an amphibious brachyuran crab that lives in the intertidal mud-flats and in mangrove swamps in the warmer parts of the world. The males have large right chelipeds that they wave in a characteristic manner. Each species has an individual 'fiddling' display differing so markedly from all others studied that closely related species could be recognized at a distance merely by the form of the display. Furthermore, related species had fundamental similarities of display in common, and series of species showing progressive specializations of structure, in general, showed a similar progression in display. W. Jacobs (1953) has described how the antenna-cleaning behaviour in some Orthoptera is characteristic of families. The Gryllidae and Tettigoniidae (and others with long antennae) clean the antennae with their maxillae; the Acrididae pull the antenna between the substratum and the tarsus whilst the Tetrigidae (Grouse-locusts) stroke their antennae with their legs and then clean their legs with the mouthparts.

The behaviour associated with nest building and the resulting structure provide important taxonomic data not only in birds but also in insects and other animals.

A. Adriaanse (1948) studied the nest-building behaviour of 69 individuals of the sandwasp, *Ammophila campestris*. These insects excavate a nest in which they place food and their egg. He found that he could divide the sample into two groups of 11 and 58 individuals respectively. Table 3 shows a comparison.

Table 3   Differences in nest building behaviour observed by Adriaanse in 1947 between two groups of *Ammophila campestris* s. lat.

	Group I (11 individuals)	Group II (58 individuals)
Breeding season	up to August	to mid-September
Food for offspring	sawfly larvae	lepidopteran larvae
Time of food provision	food added *after* egg deposited	food added *before* egg deposited
Plug formation	type I	type II

From this study came the discovery that concealed within the name *A. campestris* were two sibling species: *A. campestris* s.str. and *A. adriaansei*.

Nest structure in termites (Isoptera), which A. E. Emerson (1956) points out is the morphological expression of a behaviour pattern, has been intensively studied. Infraspecific nest building behaviour patterns have been investigated by K. Gösswald (1941) in different races of the ant *Formica rufa*.

The Paridae (titmice) are an apparently fairly homogeneous group of passerine birds but a study of their behaviour, particularly in relation to nest-building suggests that they may be an artificial assemblage of polyphyletic origin. The true titmice (*Parus* and its relatives) nest in hollow trees and other crevices; the bushtits (*Psaltriparus* and its relatives) nest in bushes and trees producing an oval-shaped structure with a lateral entrance – this is a particularly homogeneous group of social birds with similar habits and song; the penduline titmice (*Remiz* and its relatives) form very characteristic retort-shaped nests of plant down worked into a material of felt-like consistency; the bearded tits (*Panurus*), although they share many features with the Paridae, stand somewhat apart from the rest and show some affinity with other passerine families. For example, certain features of their internal morphology appear to relate them to the finches (Fringillidae) whilst in their behaviour they show similarities with the babblers (Timeliidae). The latter is itself, in all probability, a polyphyletic group. The distinctness of *Panurus* from other parids and the conflicting evidence as to their true taxonomic position has led other authors to separate them as a distinct family, the Panuridae.

In the avian order Charadriiformes (sandpipers, plovers and related birds) some doubt has existed concerning the true systematic position of certain genera, e.g. *Haematopus* (oystercatchers), *Himantopus* (stilts) and *Recurvirostra* (avocets). The stilts and avocets show a clear affinity to one another and are generally placed in the same family the Recurvirostridae, which itself is often allied to the sandpipers, curlews and godwits (Scolopacidae). The oystercatchers, on the other hand, are usually placed in a family of their own, the Haematopodidae, which is undoubtedly close to the Charadriidae (plovers and lapwings). Studies on the behaviour, especially with regard to head-scratching, in the three genera has shown that they should perhaps all be placed close to one another and to the Charadriidae rather than the Scolopacidae. These findings are supported by other biological and by some anatomical features.

The existence of conflicting evidence is very common in many taxonomic situations. This frequently arises from convergence. An interesting case of behavioural convergence is exhibited by the African horned-viper (*Cerastes cerastes*) and the American rattlesnake (*Crotalus cerastes*), the so-called sidewinder rattlesnake. These snakes belong to separate families, Viperidae (true vipers) and Crotalidae (pit vipers and rattlesnakes) respectively. Each has separately evolved the method of locomotion known as 'sidewinding' (Fig. 5–4) which is an adaptation for movement over loose sand. They also exhibit other adaptive convergences both in structure and colour.

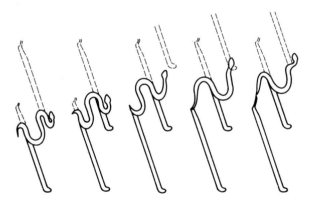

Fig. 5–4   *Cerastes cerastes*, sidewinding progression. (From Mosauer, W., 1932, *Science*, **76**, 583–5.)

Mention has been made of the use of nest building behaviour in termites, ants and birds. Specific behaviour patterns result in the creation of nests of characteristic form and composition. Constructions, such as nests, are termed in taxonomy the 'works' of an animal. Under this heading there are also many other kinds of structure which can be used to provide taxonomic characters and are even, under certain circumstances, admissible in nomenclature. Works of animals include features such as tracks and trails, insect and mite induced galls, patterns produced by bark beetles, wood, rock and shell borers, and by larvae of leaf-mining insects. Many marine polychaetes (bristle worms) produce tubes, some are of a fragile and transient nature only but others that are calcareous and persistent.

These features all provide useful taxonomic characters that are frequently easier to observe and distinguish one from another than those derived from the animals themselves. The gall-producing mites of the family Eriophyidae are very minute and usually difficult to identify. They are, however, very host-specific and frequently produce quite characteristic and easily recognizable galls (Fig. 5–5).

Associations of another kind between different groups of living organisms that are useful in taxonomic studies are described in the next chapter.

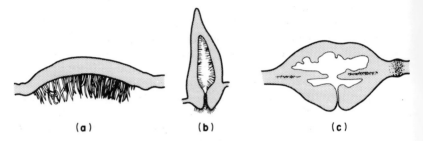

(a)   (b)   (c)

Fig. 5–5 Sections of three kinds of gall produced by eriophyid mites: (a) felt gall or erineum on a vine leaf, (b) a nail gall on lime and (c) a blister gall on walnut. (From Evans, G. O., Sheals, J. G. and Macfarlane, D., 1961, *The Terrestrial Acari of the British Isles*. Brit. Mus. (Nat. Hist.), London, p. 123; based on Ross.)

6 Host-Parasite and Host-Symbiont Relationships in Taxonomy

Considerable evidence exists that following divergent evolution or radiation of a host ancestor there ensues a corresponding speciation of its parasites and/or symbionts. This, it is believed, gave rise to the phenomenon known as *host specificity*, i.e. the restriction of a parasite or symbiont to a particular host species. Such parasites or symbionts are said to be *stenadaptive* or *monoxenous* (Fig. 6–1a) as opposed to *euryadaptive, oligo-* or *polyxenous* (Fig. 6–1b) organisms. If this is correct, then, in general, closely-related hosts should have closely-related parasites and symbionts, and it should be possible to use them as indicators of the taxonomic affinity of their hosts.

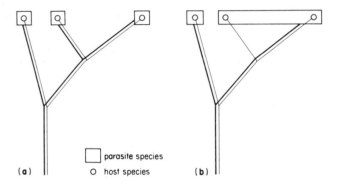

☐ parasite species
○ host species

(a)   (b)

Fig. 6–1 Diagrammatic representation of (**a**) stenadaptive and (**b**) euryadaptive parasites.

Following on from this proposition there are three so-called parasitophyletic rules:

1. *Fahrenholz's Rule*   the phylogenetic relationships of parasites reflect those of their hosts.

2. *Szidat's Rule*   the more primitive hosts have more primitive parasites, more advanced hosts have correspondingly more advanced parasites.

3. *Eichler's Rule*   taxa with much systematic variation and many sub-taxa have a greater number of species of parasites than is found in a more uniform group.

These are, at most, generalizations. We have already seen (p. 18) in the case of the flamingos an example where Mallophaga and Cestoda are apparently poor indicators of the taxonomic position of host birds. The flamingos have probably gained their mallophagan parasites secondarily, and this has almost certainly occurred fairly recently from geese by virtue of their sharing a common aquatic habitat. They may have derived their cestodes from the charadriiform waders via a common food species that acted as a secondary host.

Animals appear sometimes to derive their ectoparasites from others on which they feed, for example some predatory birds are known to share mallophagan parasites with their prey. Similarly at least some Siphunculata (sucking lice) ectoparasitic on carnivorous mammals are believed to have been obtained from rodent prey.

In spite of such cases, there are now a number of instances known where host-parasite relationships are sufficiently well correlated to regard the use of such data as significant even if not absolutely reliable.

Although it is more usual for parasites to act as indicators of the taxonomic position of their host, the reverse is also possible, at least to some extent.

The environment of an internal parasite varies far less than that possible for its host, whereas the host's true relationships may be masked by external morphological adaptive change, their internal parasites probably remain comparatively unaltered. Ectoparasites are more liable to adaptive modification in association with morphological or environmental changes of the host.

The principle involved in the use of parasites as indicators of their host's affinities has been termed by Baer the *Principle of Phylogenetic Specificity.*

Existing evidence suggests that the Mallophaga became parasitic on birds early in the history of the latter and that they evolved together. On the whole bird lice are highly specific to their hosts. This is especially true of those living on the bird's head. The specificity is so great that occasional accidental interspecific transfers very frequently result in the death of the parasite. The specificity in the Mallophaga appears to be due in part to the feather structure of the host, but there is also evidence that some factor in the chemical composition of the feathers, skin or blood may also be of importance.

On the basis of this high specificity in bird lice T. Clay (1957) suggested a classification of birds that is similar to that of A. C. Chandler (1916) based on his comparative study of feathers. It is, however, different in many respects from the classification of recent birds proposed by E. Mayr and D. Amadon (1951).

The South African ostriches (*Struthio*) and the South American *Rhea* are both parasitized by Mallophaga of the genus *Struthiolipeurus*. This genus is found on no other birds and the two species *S. struthionis* and *S. rheae* are very similar and undoubtedly closely related. On this evidence, Clay has suggested a common origin of the two orders of flightless birds, the Struthioniformes and the Rheiformes.

In addition to the shared Mallophaga, ostriches and rheas have a common tapeworm (*Houttuynia struthiocameli*), a mite (*Pterolichus bicaudatus*) and a genus of threadworm (*Dicheilonema*) – with *D. spicularia* in the ostrich and *D. rheae* in the rhea.

There are, of course, also some morphological grounds for uniting these two orders of birds. In spite of all this evidence, a monophyletic origin for the Struthioniformes and the Rheiformes has been disputed on zoogeographical grounds.

It has been pointed out by the protagonists of a diphyletic origin that Africa and South America have been separated since the break up of Pangaea in the Mesozoic era and that the earliest fossils of these birds did not appear until the Tertiary Period, perhaps some 50 million or more years later. This does not, however, preclude their origin from a common avian ancestor. It is always possible, of course, that fossils existed in pre-Tertiary times but have not yet been discovered.

Other examples of host specificity are found in the parasitic association between some fish and their tapeworms and flukes (monogenetic trematodes) and also in the occurrence of sucking lice on humans and some other primates.

A different aspect of host-parasite relationships in taxonomy is the discovery of sibling species of hosts – as a result of parasite differences. For example *Octopus bimaculatus* was found to have two species of mesozoan parasites: *Dicyemennea abelis* and *D. californica*. It was eventually noted that those infested with *D. abelis* differed in habitat and in egg structure from those with *D. californica*. The two groups of octopus were subsequently found to be reproductively isolated and are now regarded as sibling species.

Host specificity is found in symbiotic as well as parasitic relationships. For instance, there are now many examples known of the restriction of groups of related flagellate Protozoa to groups of related termites. It is often possible to identify the genus, or at least the family of the host by its flagellate fauna. In some cases identification, in this way, is even possible to the species. Furthermore, phylogenetic speculations in both the flagellates and their termite hosts seem to coincide.

Detailed studies of the homopteran bugs of the superfamilies Fulgaroidea (lantern-flies), Jassoidea (leaf-hoppers), Cercopoidea (frog-hoppers) and Cicadoidea (cicadas) and their symbionts notably by H. J. Müller, Ermisch and Wagner have shown that there is a strong correlation between the members of these superfamilies and their symbionts.

These authors derive the four superfamilies (all belonging to the suborder Auchenorrhyncha) from the suborder Coleorrhyncha. Their symbionts can

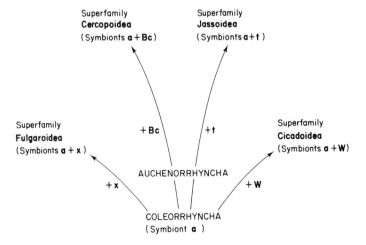

Fig. 6–2 Supplementation of ancestral symbiont a with x, Bc, t, or W in the four descendent superfamilies.

similarly be derived by appropriate supplementation (Fig. 6–2).

Within each of the four subfamilies further additions and deletions occur. In the Fulgaroidea, for example, the subfamilies of the Delphacidae exhibit the alterations shown in Fig. 6–3.

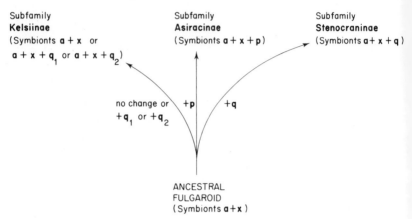

Subfamily
Kelsiinae
(Symbionts **a** + **x**   or
a + **x** + **q**$_1$ or **a** + **x** + **q**$_2$)

Subfamily
Asiracinae
(Symbionts **a** + **x** + **p**)

Subfamily
Stenocraninae
(Symbionts **a** + **x** + **q**)

no change or
+**q**$_1$ or +**q**$_2$

+**p**

+**q**

ANCESTRAL
FULGAROID
(Symbionts **a** + **x**)

Fig. 6–3   Probable evolutionary sequences of symbionts in some subfamilies of delphacid fulgaroids.

Extensive studies have also been carried out with scale insects (Coccoidea) and their symbionts.

Interspecific relationships of kinds other than those between parasites, symbionts and their hosts may also be of some taxonomic value.

The existence of sibling species of termites has been revealed after the discovery of different species of beetles of the family Staphylinidae that habitually live in the termite colonies.

Another kind of interspecific relationship is that between phytophagous organisms and their plant hosts. There is very little difference between an ectoparasite and, for example, an aphid feeding on a host plant.

R. Blackman (1974) has drawn attention to the usefulness of knowing the host plant when identifying closely-related species of aphids which frequently occur on different plant species. Aphids are not very host specific and several species may be found on the same plant. In these cases however, they are generally only distantly-related to one another.

When utilizing host plants as taxonomic characters, it is necessary to bear in mind the possible occurrence of phenotypic plasticity. This may result in morphological changes in individuals of the same species when found on different plant hosts or even on different parts of the same plant. The scale insect *Diaspidiotus ancylus*, for example, lives on deciduous trees. It may occur on leaves or bark depending on the time of year. The leaf form is different structurally from the bark form into which it transforms after leaf-fall. The change is so marked that these individuals were at one time regarded as a distinct species and genus, *Hemiberlesia howardi*. Striking ecophenotypic variations such as this are not uncommon.

7 Cytotaxonomy

Cytology and cytogenetics have had a considerable influence on the development of plant taxonomy (HEYWOOD, 1976), but a much smaller one in the case of animals. Cytotaxonomy covers all aspects of taxonomy at a cellular level, including structural, genetical, physiological and biochemical data. Phenomena associated with the nucleus are are also included, these come under the heading of karyology. Data from all of these sources can be used as systematic criteria in the same way and with the same reservations as those from the other sources discussed elsewhere in this book.

Most cytotaxonomic work so far has been concerned with the number, morphology and behaviour of chromosomes, but cytochemical information is now also being used.

7.1 Chromosome number

Chromosome numbers vary in animals from a diploid number of 2 in *Parascaris equorum* of the variety *univalens* to one of 446 in a North African lycaenid butterfly (*Lysandra atlantica*). In the majority of animals, however, the number varies from 12 to 60. Considerable variation even occurs within single orders or families. In Diptera (two-winged flies) for example, the diploid number varies from 4 to 20 and in Lepidoptera from 14 to 446.

Chromosome numbers are not necessarily the same even in related genera. For example, in the marsupial family Didelphidae (opossums) the diploid number may be 18, 22 or 28. In this family the number appears to be fairly constant within a genus, e.g. species of *Monodelphis* have 18 and those of *Didelphis* have 22. Variations may occur within a single polytypic species. In the marine isopod crustacean, *Jaera syei*, there are six races which have diploid numbers in males of 18, 20, 22, 24, 26 and 28.

The most commonly occurring number within a group does not always coincide with what is normally considered to be the ancestral form, although this is often apparently true.

Polyploidy in animals is rare. Among hermaphrodite animals autotetraploids occur, for example, in Turbellaria (free-living flatworms), Oligochaeta (earthworms), which have many polyploid races and species, and in Basommatophora (freshwater snails) among which octoploidy occurs.

Autopolyploids also occur in some parthenogenetic annelids, crustaceans, insects and vertebrates. Somatic haploidy may occur in parthenogenetic animals, but in some, for example certain parthenogenetic grasshoppers, a doubling or diploidization of the haploid egg chromosomes may take place.

In some sexually reproducing animals, e.g. some amphibia and reptiles, allotetraploids are found. In the South American family of toads, Ceratophrydidae, diploids, tetraploids and octoploids all occur, and in *Cnemidophorus*, the whiptail lizard of the Arizona Desert, diploids, triploids and tetraploids are found.

A different kind of chromosome number change sometimes occurs. During reduction division (see KEMP, 1970), the individuals of a homologous pair of chromosomes may fail to separate (non-disjunction). Fertilization of the gamete bearing a homologous pair will give rise to a zygote with three instead of the normal pair of chromosomes. Cases of this sort are in general known as aneuploids and in this particular example as trisomics. A well-known case that occurs in one or two out of every thousand human births is that of Down's syndrome popularly known as mongolism in which chromosome 21 is a trisomic (Fig. 7–1). Aneuploidy occurs only rarely in animals and is of no significance in taxonomy. It is, however, a situation that one must look out for in karyotype examination. Aneuploidy is more common in plants, especially in polyploids where its occurrence may be of taxonomic importance.

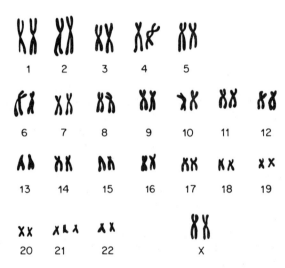

Fig. 7–1 Idiogram of human female with Down's syndrome showing trisomic aneuploidy for chromosome 21. Compare with Fig. 7–6. (From Strickberger, M. W., 1968, *Genetics.* MacMillan, New York.)

An aspect of chromosome number change that is important in animal taxonomy is that of the supernumerary or B-chromosome. The origin of these is usually unknown. The presence of B-chromosomes seems to have little or no effect on the morphology of the animal and are consequently more common than monosomics or polysomics. Supernumerary chromosome occurrence varies from one population of a species to another, both in number and frequency. For a fuller description of this and other chromosome variations see JOHN (1976).

7.2 Chromosome structure

In genera where species are found to have the same number of chromosomes, differences in their structure frequently occur, for example, in the position of the centromere (the point of attachment of the chromosome to the spindle). Three main types of spindle attachment are usually recognizable. These are referred to as metacentric, acrocentric and telocentric depending on the position of the centromere (Fig. 7–2). Sub-divisions of these are also recognizable.

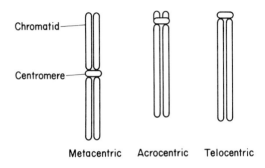

Metacentric   Acrocentric   Telocentric

Fig. 7–2 Spindle attachment.

Differences of centromere position are present, for example, in the didelphid *Marmosa*. In this genus the X-chromosome may be metacentric (*M. robinsoni*), submetacentric (*M. murina*) or acrocentric (*M. alstoni*).

At the other extreme to monocentry is holocentry in which the chromosomes are attached along their entire length to the spindle. This occurs in some scorpions, homopteran and heteropteran bugs and in the Lepidoptera.

Another type of structural variation is that resulting from translocation, arising either from breaks in two or more non-homologous chromosomes with incorrect rejoining. Most commonly this takes the form of a mutual change between the chromosomes, this is known as a reciprocal translocation (Fig. 7–3). It is often difficult to distinguish a reciprocal translocation from a simple translocation or relocation, either terminal or non-terminal. In both these cases a section of the chromosome breaks away and is attached to another without a reciprocal interchange.

Chromosome fusions are also known to take place as, for example, in the tobacco mouse (*Mus poschiavinus*) where a diploid number of 26 has been obtained by seven different fusions of the 40 chromosomes of the ancestral house mouse (*M. musculus*).

Following chromosome breakages, without relocation, the isolated fragment may be lost as it usually lacks a centromere. Such deletions tend to be lethal. If a break occurs at both ends of the same chromosome, a ring-chromosome is sometimes formed by a joining up of the broken ends. These are nearly always non-heritable.

A common type of chromosome mutation involves two breaks followed by repair but with the median segment turned upside down (Fig. 7–4). These are

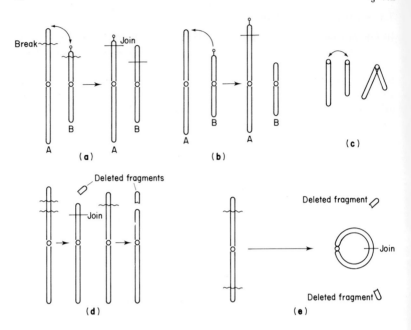

Fig. 7–3   Chromosome mutations **(a)** reciprocal translocation **(b)** relocation **(c)** fusion **(d)** deletion **(e)** ring chromosome formation.

known as inversions. They are not usually obvious except in giant polytene chromosomes such as are found in the salivary glands of Diptera, some Collembola (primitive wingless insects known as springtails) and a few ciliate Protozoa (meganucleus only). Inversions may be paracentric or pericentric.

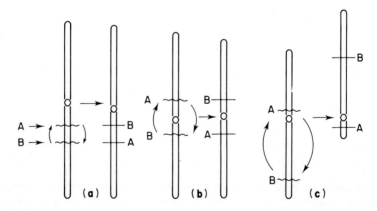

Fig. 7–4   Chromosome inversions **(a)** paracentric **(b)** pericentric without change in position of the centromere **(c)** pericentric inversion with change from meta- to acrocentric condition.

Inversions of the latter type may result in a change from acro- to metacentric chromosomes. This has happened for example in some genera of grasshoppers, and in *Drosophila* the karyotype of *D. ananassae* differs from that of *D. melanogaster* in that a pericentric inversion in the X chromosome has occurred in the former species to convert it from an acrocentric (see Fig.7–2) to a metacentric condition. In *Drosophila* the most frequent interspecific chromosome differences are those due to paracentric inversions. Stone has estimated that the average species of *Drosophila* is polymorphic for 14 paracentric inversions and perhaps some 70 000 rearrangements of this type have occurred in the 2000 or so species in the genus.

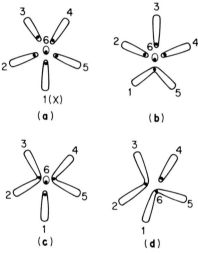

Fig. 7–5 Male haploid karyotypes of some species of *Drosophila*: **(a)** *D. subobscura* showing probable primitive number of six acrocentric chromosomes. **(b)** *D. pseudoobscura* with a large metacentric chromosome formed by the fusion of 1 and 5. **(c)** *D. melanogaster* with fusions between 2 and 3, and between 4 and 5. **(d)** *D. willistoni* with fusions between 2 and 3, and 1, 6 and 5.

Some chromosomes possess small chromatin bodies, known as satellites attached by threads to the end of the chromosome (see Fig. 7–3 (a) and (b)).

In addition to the well-known banding on the giant salivary gland chromosomes, particular regions of chromosomes stain differentially with Giemsa stain and quinacrine derivatives. These produce G- and Q-bands respectively and give each chromosome its own particular banding pattern. These are very helpful in the identification of chromosomes of similar shape and size, and in comparisons between species or races.

J. T. Patterson and W. S. Stone (1952) have analysed the cytotaxonomic differences between 200 species of *Drosophila* based on somatic metaphase configurations and on polytene salivary gland chromosomes. Some quite marked differences were found in the structure and number of chromosomes (Fig. 7–5). The basic primitive number is probably 6. In some species such as

D. mulleri, D. aldrichi and *D. wheeleri* no differences were found, even in the polytene chromosomes.

Comparisons of chromosomes are usually made between the metaphase configurations. These are known as karyotypes (Fig. 7–5). When the number of chromosomes is large they are often arranged in decreasing order of size, such pictorial arrangements are termed idiograms (Figs 7–1 and 7–6).

Chromosomes are usually numbered in order of size from the largest (1) downwards (as shown in Fig. 7–6), according to their arrangement in the idiogram.

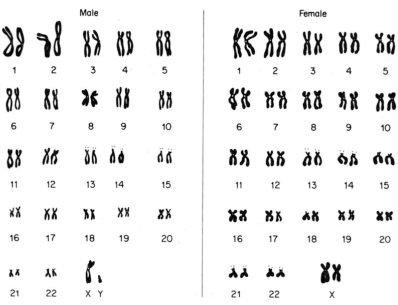

Fig. 7–6   Idiogram of normal human chromosomes. Compare with Fig. 7–1. (From McKusick.)

Chromosomes of similar overall length can be distinguished by one or more of the following: (*a*) position of the centromere, (*b*) the *p*/*q* ratio, i.e. that between the length of the arms on either side of the centromere, (*c*) presence or absence of satellites, and (*d*) the Q- and G-banding patterns.

Inspection of the karyotype has sometimes revealed the existence of sibling species concealed under a single name. For example, *Drosophila persimilis* and *D. pseudoobscura* were believed to be conspecific until their karyotypes were compared. Their distinct nature has since been confirmed by other characters such as those of the male genitalia. The species and subspecies of the *Anopheles maculipennis* complex (see p. 2) are now known to have different karyotypes.

Karyotype data have not proved to be of use in taxonomic discrimination in all groups of animals. In many genera there appear to be few or no easily demonstrable differences in either chromosome number or morphology between species. For example, among the grasshoppers there is great uniformity in the

Acrididae ($2n\,\male = 23$ acrocentrics), Pamphagidae ($2n\,\male = 19$ acrocentrics) and Pyrgomorphidae ($2n\male = 19$ acrocentrics). In the archaic family of grasshoppers, Eumastacidae, $2n\male = 13\text{--}21$. Even in groups such as the acridoids where there is great stability, karyotype data may still be of some use. The New World grasshoppers of the genus *Phrynotettix* have been considered by some entomologists to be related to the subfamily Romaleinae and by others to be the only American representatives of the Pamphagidae. *Phrynotettix* has a diploid number in males of 23 acrocentrics as in the Romaleinae. The corresponding number in species of Pamphagidae is 19.

Chromosome studies, like those of chemo- and immunotaxonomy, have an important part to play in the gamma phase of taxonomy. Various kinds of intraspecific karyotype variations are now known to occur. Polymorphism in pericentric inversion in one of the chromosomes of the black rat *Rattus rattus* has been studied in detail in Japan by T. H. Yosida, K. Tsuchiya and K. Morikawa (1971). The same authors have made a careful study of chromosome variation in *R. rattus* in South-East Asia, where the species probably first evolved, and elsewhere. They confirmed the existence of some subspecies and recognized others on the basis of their chromosomes, especially in respect of pericentric inversions and of fusions.

Geographical races (subspecies) are also known, for example, in the European and Canadian races of the midge, *Chironomus tentans*. The difference is recognizable in the banding patterns of their polytene (giant) salivary gland chromosomes.

Another example is found in the greater sand gerbil, *Gerbillus pyramidum*, in which there are three distinct geographical races each having a different diploid number: the Algerian race $2n = 40$, the Israel coastal race $2n = 52$ and the Negev and Sinai race $2n = 66$. Hybrids with diploid numbers varying from 51 to 61 occur between the coastal and Negev races. The hybrid karyotypes exhibit clines not only in chromosome numbers, but also in the proportion of metacentrics.

Karyological studies have revealed the existence of other clinal variations in chromosomal features related, as in the case of *Clupea pallasi* (see p. 7), to environmental gradations. In the dog-whelk, *Nucella lapillus*, common round our coasts in the middle shore on rocky beaches, the diploid chromosome number varies from 26 to 36. The lowest chromosome number is found where wave and general tidal movement is greatest. An increase is observed on more sheltered beaches. Clines are also known in the percentage occurrence of different inversions in the Neotropical *Drosophila flavopilosa*.

In using karyological data in taxonomy, it is important to be conscious not only of geographical races, individual abnormalities, clines and other kinds of polymorphism, but also of seasonal differences. These have been found, for example by J. Wahrman (1954) in Israel in the mantid *Ameles heldreichi*. In this case the males may have a diploid number of chromosomes of 27, 28 or 29. There are two generations a year, one in spring and the other in the autumn. Among these $2n = 27$, for example, is significantly more common in the second than in the first generation. In this as well as in the clinal variations studied there must be some adaptive advantage maintaining the differences.

8 Numerical Taxonomy

The use of numerical methods in taxonomy is not new. Simple statistical techniques like those involving standard deviations, t tests and chi-squared have been employed for many years. Details of these and other methods are given in many elementary texts such as PARKER's *Introductory Statistics for Biology* in this series and more comprehensively in SIMPSON, ROE and LEWONTIN's *Quantitative Zoology*. No further reference will be made to them here. With the advent of electronic digital computers a new branch of taxonomy known as numerical taxonomy or taxometrics has developed in the last twenty years. Computers have permitted analyses of large quantities of data in relatively short spaces of time. The computations themselves probably take a few seconds only, but the preparation of the data still requires painstaking examination and recording of information in a form suitable for input to the computer.

By these means large numbers of characters from many organisms can be compared with one another. Groupings according to overall similarity or dissimilarity can be made and, where necessary, presented graphically. In practice about 50 to 100 characters are employed, usually from approximately the same or a greater number of organisms. Useful information can, however, still be obtained by these methods with much smaller numbers of animals and/or characters. The use of the word 'organisms' here can refer to individuals, populations, species, genera or any other taxonomic category, and for this reason they are generally referred to as *operational taxonomic units* or more usually and simply as OTUs.

The different conditions in which the characters occur are known as *character states*. A particular organ may, for example, be present or absent, or black or white. In such simple cases they are said to be *two-state characters*. Many, however, exhibit a number of possible states – these are termed *multi-state characters*. Characters of either sort may be qualitative or quantitative.

The first step in numerical taxonomy after deciding upon the OTUs to be studied is to select the characters to be employed. In general the larger the number of characters, the better will be the result. Characters need not all be derived from internal or external morphology, they may include any attributes of the OTU (biochemical, developmental, ecological, behavioural etc.) In all cases, however, they are observable phenotypic characters. For this reason any grouping, classification or key that results from taxometric methods is said to be *phenetic* and is entirely without phylogenetic or *phyletic* pretensions. In the absence of detailed palaeontological evidence, they probably approximate as closely as is possible to a truly phyletic one. It is precisely because they are based entirely on observable similarities or dissimilarities, without any regard to supposed evolution, ancestry or genetic relationship that some systematists regard numerical taxonomy as a useful tool but not as providing a satisfactory so-called 'natural' classification.

In numerical taxonomy characters are not generally given any *a priori* weighting – each is of the same value as any other. There are, however, some inadmissible characters as mentioned on p. 7. Numerical methods are available to weight characters *a priori* or on some empirical basis after their distribution among the OTUs has been analysed.

The character states must next be recorded in a suitable form. This process is known as *coding*. Quantitative attributes such as scale-counts in reptiles or measurements of parts of the body can be recorded directly. Two-state or multi-state characters are coded. In the case of two-state characters presence is usually coded as + or 1, and absence as – or 0.

Where a character exists in more than one qualitative state, these may be broken down into a series of two-state characters. If, for example, the cuticle of an insect may be white, yellow, brown or black, the character can be coded for possession (+ or 1) or absence (– or 0) of each state in turn. Alternatively, each of the possible states can be indicated, usually by a number, and agreement or disagreement (usually termed match or mis-match) be recorded as + or – for each comparison between OTUs.

In the case of quantitative multi-state characters of either the continuous or meristic (see p. 7) kind, the states may be ranked in order of magnitude and coded 1, 2, 3, etc. For this reason such characters are sometimes termed *ordered*

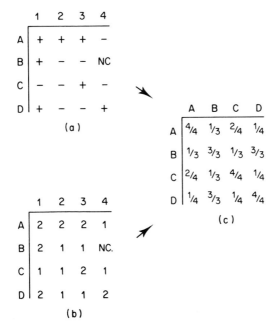

Fig. 8–1 A primary data matrix based on two-state characters, expressed in two ways (**a**) and (**b**); (**c**) method of calculating the coefficient of similarity.

multi-state characters as opposed to unordered qualitative characters. Quantitative multi-state characters of the continuous kind can also be grouped into a series of two-state characters, e.g. setae less than 5, or setae 5 or more; setae 5–8 or setae 9 or more, etc. Sometimes, for one reason or another, no comparison is possible or data may be missing for one of the OTUs. Situations of this kind are coded as NC (no comparison).

After selection of OTUs, the determination of their character states and their subsequent coding, a *data matrix* or $t \times n$ matrix can be constructed, n representing the characters and t the OTUs as shown in Fig. 8–1. (a) and (b).

The next stage is normally the comparison of each OTU in turn with all others in respect of their character states. To do this some means of comparing the degree of similarity must be employed. In the example shown in Fig. 8–1, where two-state characters are employed, this can be achieved simply by noting what is called the *coefficient of similarity* (S). $S = \frac{m}{n}$ where m is the number of matches of character states between pairs of OTUs, and n is the total number of characters (NC entries are excluded). The coefficients of similarity for the OTUs shown in Figs 8–1 (a) and (b) are indicated in (c) as fractions. It is more usual to express these as decimal fractions or percentages. In the latter case 0 would indicate total dissimilarity and 100 would represent phenetic identity. This method is used in the example illustrated in Figs 8–2, 8–3 and 8–4. These represent the simplest of a large number of possible techniques and is known as *single linkage cluster analysis* in which the measure of similarity is based on match/mis-match or one of the many other methods.

The presentation of information shown in Figs 8–1(c) and 8–2 is in the form of a *similarity matrix*, i.e. a tabular comparison of each OTU with every other one.

	A	B	C	D	E	F	G	H	I	J
A	100	54	80	63	62	81	50	83	50	61
B	54	100	55	57	57	55	86	56	87	56
C	80	55	100	62	64	85	51	86	50	62
D	63	57	62	100	74	63	56	65	56	96
E	62	57	64	74	100	64	56	67	56	72
F	81	55	85	63	64	100	54	87	52	65
G	50	86	51	56	56	54	100	54	85	55
H	83	56	86	65	67	87	54	100	54	67
I	50	87	50	56	56	52	85	54	100	55
J	61	56	62	96	72	65	55	67	55	100

Fig. 8–2 A similarity matrix.

It will be noted that the figures on the upper and lower sides of the diagonal line in Fig. 8–2 are mirror images. Consequently it is customary to illustrate one part only as in Fig. 8–3.

The OTUs are next grouped into *clusters* on the basis of their similarity. Again there are many techniques available for doing this. With the small numbers of OTUs used in the example shown in Figs 8–2 – 8–4 simple visual inspection will immediately show a high degree of similarity between A, C, F and H, between D, E and J and again between B, G and I. The matrix can now be rearranged or 'organized' to form blocks of high similarity as shown in Fig. 8–3.

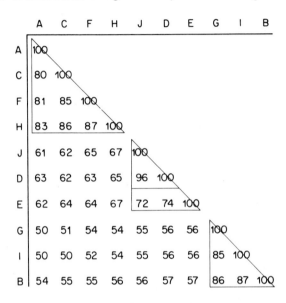

Fig. 8–3   The similarity matrix rearranged to show clusters of OTUs.

When the matrix is very large the rearrangement is more difficult and time-consuming. In these cases the data can be punched on cards and computer aided techniques can be employed. There is now a considerable number of computer programs available for the numerical taxonomist.

The percentage of shared character states that forms the basis of comparison of OTUs in the example shown also permits graphical representation to be made which can take the form of a kind of tree or dendrogram known as a *phenogram*. With this type of data presentation, phenetically similar OTUs can be linked together by horizontal lines placed at the appropriate height in relation to a vertical scale representing degrees of similarity (in this case expressed as a percentage). The similarity shown in Fig. 8–3 can thus be indicated in the phenogram (Fig. 8–4). Clusters of similar OTUs that are clearly evident are known as *phenons* and the levels indicating the degrees of similarity on the vertical scale are termed *phenon lines*. In the hypothetical dendrogram illustrated, OTUs A, C, F and H form a group or phenon having 80–87%

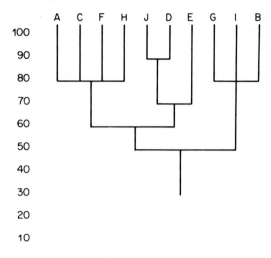

Fig. 8–4   A phenogram based on the information shown in Fig. 8–3.

similarity. D and J have 96% similarity, and are both linked to E at about the 70% level. G, I and B are linked at the 85–87% level. The group formed by J, D and E is joined to A, C, F and H at the 60% phenon line. More remote from all of these seven are G, I and B to which they are linked with only 50–57% similarity as shown in Fig. 8–4.

It is very tempting but not justifiable to regard such dendrograms as phylogenetic trees or, at least, approximations to them. Partly for this reason, and partly because of the subjective nature of categories such as genus, subfamily, family etc., levels of phenon lines have been used to delimit taxa above the species level. For example, phena with about 70% similarity might be regarded as of generic rank, and those of say 50% of family rank. This provides an explicit but arbitrary method of standardizing the level of various categories in the taxonomic hierarchy. Whether or not all genera and families, for example, should be delimited in this way by more-or-less constant levels of phenon lines is a debatable point.

Many other techniques of numerical taxonomy are available, some with special objectives. OTUs may be depicted as clusters of points in two- or three-dimensional diagrams, or methods can be used which will maximize the difference between known taxonomic groups and so facilitate their discrimination. Other methods can be used to compare alternative classifications of the same set of OTUs, based on different taxonomic criteria or different character sets. Intensive studies of the numerical properties of characters and the relations between them (*character analysis*) can also be made and subsequently used in more orthodox taxonomic discussions. Numerical techniques are particularly appropriate for handling quantitative data obtained by biochemical, serological and genetic methods and not readily amenable to discussion in terms of the qualitative distinctions commonly encountered in classical taxonomy.

Classifications based on the methods of numerical taxonomy almost certainly have some phyletic component because they result, at least to some extent, from similarities due to common ancestry. Similarity of this kind is said to be *patristic* as opposed to those which result from convergent evolution in more-or-less unrelated taxa. This kind of similarity is termed *homoplastic*.

The concept of character weighting was referred to briefly in Chapter 2 where it was stated that a 'good' character (i.e. one of high weight) should, according to many taxonomists, be given special consideration in the construction of classifications. The determination of the 'goodness' of a character has tradition-ally been made on the basis of a number of observations during the early stages of the beta phase of taxonomic studies. Such considerations are those of consistency of occurrence, genetic stability, lack of too much individual phenotypic or genotypic variation, those which, on the basis of previous studies, are believed to be of phylogenetic significance or to have been found to be of value in studies on related taxa. To give special weight to such characters in the construction of a classification is said to give *a priori* weighting. Some taxonomists maintain that characters such as those provided by immunological or DNA studies should be given high weighting and used *a priori* in the construction of classifications and phylogenies. *A priori* weighting is rarely employed in numerical taxonomy. However, once the procedures outlined above have been completed and a classification constructed, it is then possible to reconsider the characters employed and to determine which are constant and highly correlated with other characters, i.e. which are 'good'. These characters are given special weighting in diagnosing taxa and identifying them by keys. Such weighting is said to be *a posteriori* or *correlation* weighting.

Following the development of computer techniques in numerical taxonomy, there have been other applications such as those used in key construction and automatic identification programs (PANKHURST, 1978), in storage and retrieval of taxonomic data and in studies on evolutionary pathways.

One cannot leave the subject of taxonomy without at least a few words on a contemporary point of controversy that has arisen from a confusion between classification and phylogenetic reconstruction. For many years most biologists have believed that systems of classification should be based, as far as possible, on evolutionary considerations, and that they should aim to reflect lines of descent as well as degrees of similarity and dissimilarity. This view is still upheld by many systematists. There are some, however, who maintain that there is no way in which phenetic and phylogenetic considerations can be associated in a logical way. Within this group of taxonomists there are two main schools of thought. On the one hand there are those who base classification entirely on phenetics, i.e. on overall similarity in respect of characters (usually a large number of them) that have been given no *a priori* weighting. This method is largely supported by those who practise numerical taxonomy, but not exclusively by them – much traditional alpha and beta taxonomy is by necessity phenetic in its methods. Some of these taxonomists consider that lines of descent cannot be taken into consideration because phylogenies, in their opinion, can only be speculative hypotheses. Therefore they maintain that phenetic considerations are the only valid ones. Proponents of this view are consequently known as pheneticists or numerical pheneticists.

Many biologists have followed Simpson in regarding taxa with the maximum number of shared characters as being those which are the most closely related, and that they have diverged more recently than others with fewer shared characters. In many cases this may be true but the possibility of evolutionary convergence of characters must be borne in mind. This view cannot, therefore, be accepted, at least without serious reservations.

Others believe that the most important factor to be taken into consideration is not overall similarity but the sequence of branching. This method of classification is therefore usually termed cladistics ($\kappa\lambda\acute{\alpha}\delta o\varsigma$ = branch) but by some of its adherents, notably HENNIG, it has been referred to as phylogenetic systematics. Cladists maintain that one can determine the point of branching by careful consideration of the characters. It is necessary in this method to decide which character states are primitive (*plesiomorphic*) and which are derived (*apomorphic*). For this reason it is important to select characters for which it is possible to determine the evolutionary sequence of their states with a reasonable degree of probability. By selecting characters in this way cladistic dendrograms or cladograms can be constructed from far fewer characters than is required for numerical phenograms. Hennig made the assumption that evolution proceeds by bifurcations (*cladogenesis*) and is therefore divergent. In his system of classification chronistics or time factors are important and consequently two taxa may be placed in separate groups if the point at which they diverged was very early, even though they may have come to differ very little, i.e. *anagenesis* (degree of divergence or change) is not considered to be of importance. Hennig's original interpretation of cladistics has been modified by some subsequent systematists to form what Platnick has termed *transformed cladistics*. According to this school of workers evolution should be disregarded because, they maintain, it cannot be proved. Transformed cladism consequently relates to a purely classificatory process enabling organisms to be arranged in a hierarchical system. The requirements of both Hennigian and transformed cladistics are unacceptable to many taxonomists who prefer to adhere to the phyletic, phylogenetic or evolutionary school of systematics. Such workers not only employ phenetic and cladistic methods but also take into consideration evolutionary trends such as convergence, parallelism and anagenesis. For further discussion reference should be made to MICHENER (1970), PATTERSON (1980), CHARIG (1981) and MARTIN (1981).

The methods used in taxonomy, some of which are described in this book, are still developing rapidly and provide much of the excitement to be found in what is usually believed to be a rather dull branch of science. It is hoped that this brief account will stimulate a few experimentally minded young biologists to regard taxonomy in a different light and perhaps contribute to its resurgence as a truly scientific discipline.

Further Reading

BISBY, F. A., VAUGHAN, J. G. and WRIGHT, C. A. (1980). *Chemosystematics: Principles and Practice*. Academic Press, London.

CATCHPOLE, C. K. (1979). *Vocal Communication in Birds*. Studies in Biology no. 115. Edward Arnold, London.

CHARIG, A. (1981). Cladistics: a different point of view. *Biologist – J. Inst. Biol., London*. **28** (1), 19–20.

CROWSON, R. A. (1970). *Classification and Biology*. Heinemann Educational Books Ltd., London.

FERGUSON, A. (1980). *Biochemical Systematics and Evolution*. Blackie, London.

HENNIG, W. (1966). *Phylogenetic Systematics*. University of Illinois Press, Urbana.

HEYWOOD, V. H. (1976). *Plant Taxonomy* (2nd Edn). Studies in Biology no. 5. Edward Arnold, London.

INCHLEY, C. J. (1981). *Immunobiology*. Studies in Biology no. 128. Edward Arnold, London.

International Code of Zoological Nomenclature (1964). International Trust for Zoological Nomenclature, c/o British Museum (Natural History), London.

JEFFREY, C. (1973). *Biological Nomenclature*. Edward Arnold, London.

JOHN, B. (1976). *Population Cytogenetics*. Studies in Biology no. 70. Edward Arnold, London.

LEONE, C. A. (1964). *Taxonomic Biochemistry and Serology*. Ronald Press, New York.

KEMP, R. (1970). *Cell Division and Heredity*. Studies in Biology no. 21. Edward Arnold, London.

MARTIN, B. (1981). Phylogenetic reconstruction versus classification: the case for clear demarcation. *Biologist – J. Inst. Biol.*, London, **28** (3), 127–32.

MAYR, E. (1969). *Principles of Systematic Zoology*. McGraw-Hill Book Co., New York and London.

MAYR, E. (1970). *Populations, Species and Evolution*. Harvard University Press, Cambridge, Mass.

MICHENER, C. D. (1970). Diverse approaches to systematics. *Evolutionary Biology* (Ed. Dobzhansky, Th., Hecht M. K. and Steere W. C.) vol. 4: 1–138, New York.

PANKHURST, R. J. (1978). *Biological Identification*. Edward Arnold, London.

PATTERSON, C. (1980). Cladistics. *Biologist – J. Inst. Biol., London*, **27** (5), 234–40.

SIMPSON, G. G. (1961). *Principles of Animal Taxonomy*. Columbia University Press, New York and London.

SIMPSON, G. G., ROE, A. and LEWONTIN, R. C. (1960). *Quantitative Zoology*. Harcourt Brace, New York.

SNEATH, P. H. A. and SOKAL, R. R. (1973). *Numerical Taxonomy*. Freeman, San Francisco.

WILEY, E. O. (1981). *Phylogenetics: The Theory and Practice of Phylogenetic Systematics*. John Wiley and Sons, New York.

WRIGHT, C. A. (Ed.) (1974). *Biochemical and Immunological Taxonomy of Animals*. Academic Press, London and New York.

Index